A BOOK OF EXPOSITION

BY

HOMER HEATH HUGENT

tion et mise en page : Culturea (Hérault, 34)
rouvez notre catalogue sur http://culturea.fr/
ntact : infos@culturea.fr
primé en Allemagne par Books on Demand Gmbh
sign typographique et layout : Derek Murphy et Reedsy
3N : 9791041985623
e de parution : février 2024
livre :
a été composé avec une police open source libre de droits dessinée sur Open Fondry https://open-foundry.com/

THE EXPOSITION OF A MECHANISM

THE LEVERS OF THE HUMAN BODY

Sir Arthur Keith

In all the foregoing chapters we have been considering only the muscular engines of the human machine, counting them over and comparing their construction and their mechanism with those of the internal-combustion engine of a motor cycle. But of the levers or crank-pins through which muscular engines exert their power we have said nothing hitherto. Nor shall we get any help by now spending time on the levers of a motor cycle. We have already confessed that they are arranged in a way which is quite different from that which we find in the human machine. In the motor cycle all the levers are of that complex kind which are called wheels, and the joints at which these levers work are also circular, for the joints of a motor cycle are the surfaces between the axle and the bushes, which have to be kept constantly oiled. No, we freely admit that the systems of levers in the human machine are quite unlike those of a motor cycle. They are more simple, and it is easy to find in our bodies examples of all the three orders of levers. The joints at which bony levers meet and move on each other are very different from those we find in motor cycles. Indeed, I must confess they are not nearly so simple. And, lastly, I must not forget to mention another difference. These levers we are going to study are living—at least, are so densely inhabited by myriads of minute bone builders that we must speak of them as living. I want to lay emphasis on that fact because I did not insist enough on the living nature of muscular engines.

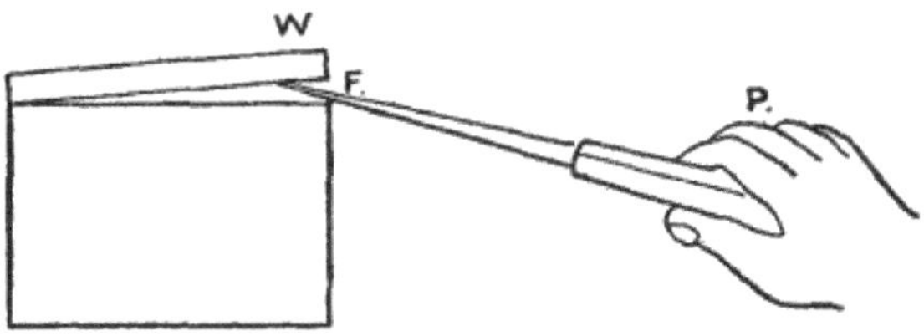

Fig. 1.—Showing a chisel 10 inches long used as a lever of the first order.

We are all well acquainted with levers. We apply them every day. A box arrives with its lid nailed down; we take a chisel, use it as a lever, pry the lid open, and see no marvel in what we have done (Fig. 1). And yet we thereby did with ease what would have been impossible for us even if we had put out the whole of our unaided strength. The use of levers is an old discovery; more than 1500 years before Christ, Englishmen, living on Salisbury Plain, applied the invention when they raised the great stones at Stonehenge and at Avebury; more than 2000 years earlier still, Egyptians employed it in raising the pyramids. Even at that time men had made great progress; they were already reaping the rewards of discoveries and inventions. But none, I am sure, surprised them more than the discovery of the lever; by its use one man could exert the strength of a hundred men. They soon observed that levers could be used in three different ways. The instance already given, the prying open of a lid by using a chisel as a lever, is an example of one way (Fig. 1); it is then used as a lever of the first order. Now in the first order, one end of the lever is applied to the point of resistance, which in the case just mentioned was the lid of the box. At the other end we apply our strength, force, or power. The edge of the box against which the chisel is worked serves as a fulcrum and lies between the handle where the power is applied and the bevelled edge which moves the resistance or weight. A pair of ordinary weighing scales also exemplifies the first order of levers. The knife edge on which the beam is balanced serves as a fulcrum; it is placed exactly in the middle of the beam, which we shall suppose to be 10 inches long. If we place a 1-lb. weight in one scale to represent the resistance to be overcome, the weight will be lifted the moment that a pound of sugar has been placed in the opposite scale—the sugar thus representing the power. If, however, we move the knife-edge or fulcrum so that it is only 1 inch from the sugar end of the beam and 9 inches from the weight end, then we find that we have to pour in 9 lb. of sugar to equalise the 1-lb. weight. The chisel used in prying open the box lid was 10 inches long; it was pushed under the lid for a distance of 1 inch, leaving 9 inches for use as a power lever. By using a lever in this way, we increased our strength ninefold. The longer we make the power arm, the nearer we push the fulcrum towards the weight or resistance end, the greater becomes our power. This we shall find is a discovery which Nature made use of many millions of years ago in fashioning the body of man and of beast. When we apply our force to the long end of a lever, we increase our power. We may also apply it, as Nature has done in our bodies, for another purpose. We have just noted that if the weight end of the beam of a pair of scales is nine times the length of the sugar end, that a 1-lb. weight will counterpoise 9 lb. of sugar. We also see that the weight scale moves at nine times the speed of the sugar scale. Now it often happens that Nature wants to increase, not the power, but the speed with which a load is lifted. In that case the "sugar scale" is placed at the long end of the beam and the "weight scale" at the short end; it then takes a 9-lb. weight to raise a single pound of sugar, but the sugar scale moves with nine times the speed of the weight scale. Nature often sacrifices power to obtain speed. The arm is used as a lever of this kind when a cricket ball is thrown.

Nothing could look less like a pair of scales than a man's head or skull, and yet when we watch how it is poised and the manner in which it is moved, we find that it, too, acts as a lever of the first order. The fulcrum on which it moves is the atlas—the first vertebra of the spine (Fig. 2). When a man stands quite erect, with the head well thrown back, the ear passages are almost directly over the fulcrum. It will be convenient to call that part of the head which is behind the ear passages the *post-fulcral,* and the part which is in front the *pre-fulcral.* Now the face is attached to the pre-fulcral part of the lever and represents the weight or load to be moved, while the muscles of the neck, which represent the power, are yoked to the post-fulcral end of the lever. The hinder part of the head serves as a crank-pin for seven pairs of neck muscles, but in Fig. 2 only the chief pair is drawn, known as the *complex* muscles. When that pair is set in action, the post-fulcral end of the head lever is tilted downwards, while the pre-fulcral end, on which the face is set, is turned upwards.

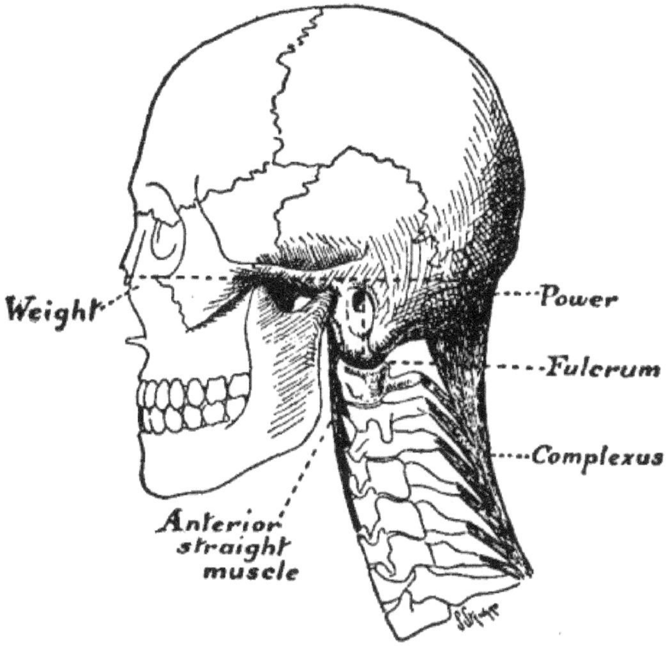

Fig. 2.—The skull as a lever of the first order.

The complex muscles thus tilt the head backwards and the face upwards, but where are the muscles which serve as their opponents or antagonists and reverse the movement? In a previous chapter it has been shown that every muscle has to work against an opponent or antagonist muscle. Here we seem to come across a defect in the human machine, for the *greater straight* muscles in the front of the neck, which serve as opposing muscles, are not only much smaller but at a further disadvantage by being yoked to the pre-fulcral end of the lever, very close to the cup on which the head rocks. However, if the *greater straight* muscles lose power by working on a very short lever, they gain, in speed; we set them quickly and easily into action when we give a nod of recognition. All the strength or power is yoked to the post-fulcral end of the head; the pre-fulcral end of its lever is poorly guarded. Japanese wrestlers know this fact very well, and seek to gain victory by pressing up the poorly guarded pre-fulcral lever of the head, thus producing a deadly lock at the fulcral joint. Indeed, it will be found that those who use the jiu-jitsu method of fighting have discovered a great deal about the construction and weaknesses of the levers of the human body.

Merely to poise the head on the atlas may seem to you as easy a matter as balancing the beam of a pair of scales on an upright support. I am now going to show that a great number of difficulties had to be overcome before our heads could be safely poised on our necks. The head had to be balanced in such a way that through the pivot or joint on which it rests a safe passageway could be secured for one of the most delicate and most important of all the parts or structures of the human machine. We have never found a good English name for this structure, so we use its clumsy Latin one—*Medulla oblongata*—or medulla for short. In the medulla are placed offices or centres which regulate the vital operations carried on by the heart and by the lungs. It has also to serve as a passageway for thousands of delicate gossamer-like nerve fibres passing from the brain, which fills the whole chamber of the skull, to the spinal cord, situated in the canal of the backbone. By means of these delicate fibres the brain dispatches messages which control the muscular engines of the limbs and trunk. Through it, too, ascend countless fibres along which messages pass from the limbs and trunk to the brain. In creating a movable joint for the head, then, a safe passage had to be obtained for the medulla—that part of the great nerve stem which joins the brain to the spinal cord. The medulla is part of the brain stem.

This was only one of the difficulties which had to be overcome. The eyes are set on the pre-fulcral lever of the head. For our safety we must be able to look in all directions—over this shoulder or that. We must also be able to turn our heads so that our ears may discover in which direction a sound is reaching us. In fashioning a fulcral joint for the head, then, two different objects had to be secured: free mobility for the head, and a safe transit for the medullary part of the brain stem. How well these objects have been attained is known to all of us, for we can move our heads in the freest manner and suffer no damage whatsoever. Indeed, so strong and perfect is the joint that damage to it is one of the most uncommon accidents of life.

Let us see, then, how this triumph in engineering has been secured. In her inventive moods Nature always hits on the simplest plan possible. In this case she adopted a ball-and-socket joint—the kind by which older astronomers mounted their telescopes. By such a joint the telescope becomes, just as the head is, a lever of the first order. The eyeglass is placed at one end of the lever, while the object-glass, which can be swept across the face of the heavens, is placed at the other or more distant end. In the human body the first vertebra of the backbone—the atlas—is trimmed to form a socket, while an adjacent part of the base of the skull is shaped to play the part of ball. The kind of joint to be used having been hit upon, the next point was to secure a safe passage for the brain stem. That, too, was worked out in the simplest fashion. The central parts of both ball and socket were cut away, or, to state the matter more exactly, were never formed. Thus a passage was obtained right through the centre of the fulcral joint of the head. The centre of the joint was selected because when a lever is set in motion the part at the fulcrum moves least, and the medulla, being placed at that point, is least exposed to disturbance when we bend our heads backwards, forwards, or from side to side. When we examine the base of the skull, all that we see of the ball of the joint are two knuckles of bone (Fig. 3, A), covered by smooth slippery cartilage or gristle, to which anatomists give the name of occipital condyles. If we were to try to complete the ball, of which they form a part, we should close up the great opening—the *foramen magnum*—which provides a passageway for the brain stem on its way to the spinal canal. All that is to be seen of the socket or cup is two hollows on the upper surface of the atlas into which the occipital condyles fit (Fig. 3, B). Merely two parts of the brim of the cup have been preserved to provide a socket for the condyles or ball.

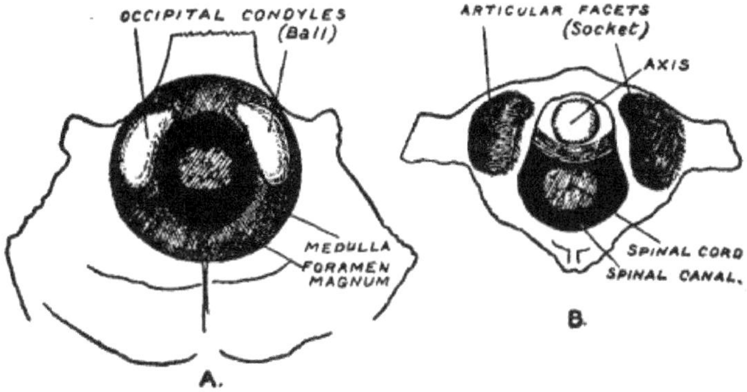

Fig. 3.—A, The opening in the base of the skull, by which the brain stem passes to the spinal canal. The two occipital condyles represent part of the ball which fits into the cup formed by the atlas. B, The parts of the socket on the ring of the atlas.

As we bend our heads, the occipital condyles revolve or glide on the sockets of the atlas. But what will happen if we roll our heads backwards to such an extent that the bony edge of the opening in the base of the skull is made to press hard against the brain stem and crush it? That, of course, would mean instant death. Such an accident has been made impossible (1) by making the opening in the base of the skull so much larger than the brain stem that in extreme movements there can be no scissors-like action; (2) the muscles which move the head on the atlas arrest all movements long before the danger-point is reached; (3) even if the muscles are caught off their guard, as they sometimes are, certain strong ligaments—fastenings of tough fibres—are so set as automatically to jam the joint before the edge of the foramen can come in contact with the brain stem.

These are only some of the devices which Nature had to contrive in order to secure a safe passageway for the brain stem. But in obtaining safety for the brain stem, the movements of the head on the atlas had to be limited to mere nodding or side-to-side bending. The movements which are so necessary to us, that of turning our heads so that we can sweep our eyes along the whole stretch of the skyline from right to left, and from left to right, were rendered impossible. This defect was also overcome in a simple manner. The joints between the first and second vertebrae—the atlas and axis—were so modified that a turning movement could take place between them instead of between the atlas and skull. When we turn or rotate our heads, the atlas, carrying the skull upon it, swings or turns on the axis. When we search for the manner in which this has been accomplished, we see again that Nature has made use of the simplest means at her disposal. When we examine a vertebra in the course of construction within an unborn animal, we see that it is really made up by the union of four parts (see Fig. 4): a central block which becomes the "body" or supporting part; a right and a left arch which enclose a passage for the spinal cord; and, lastly, a fourth part in front of the central block which becomes big and strong only in the first vertebra—the atlas. When we look at the atlas (Fig. 4), we see that it is merely a ring made up of three of the parts—the right and left arches and the fourth element,—but the body is missing. A glance at Fig. 4, B, will show what has become of the body of the atlas. It has been joined to the central block of the second vertebra—the axis—and projects upwards within the front part of the ring of the atlas, and thus forms a pivot round which rotatory movements of the head can take place. Here we have in the atlas an approach to the formation of a wheel—a wheel which has its axle or pivot placed at some distance from its centre, and therefore a complete revolution of the atlas is impossible. A battery of small muscles is attached to the lateral levers of the atlas and can swing it freely, and the head which it carries, a certain number of degrees to both right and left. The extent of the movements is limited by stout check ligaments. Thus, by the simple expedient of allowing the body of the atlas to be stolen by the axis, a pivot was obtained round which the head could be turned on a horizontal plane.

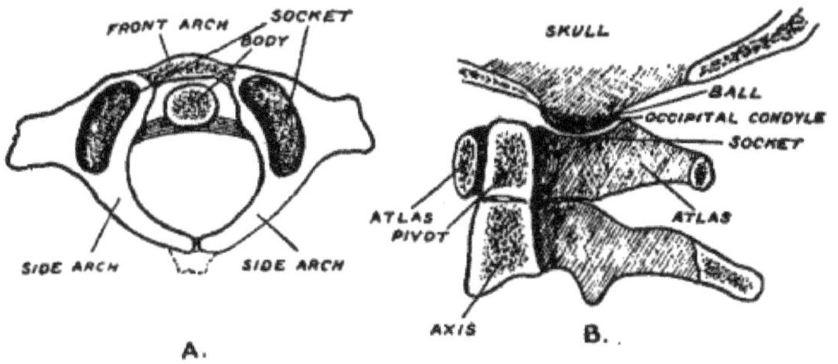

Fig. 4.—A, The original parts of the first or atlas vertebra. B, Showing the "body" of the first vertebra fixed to the second, thus forming the pivot on which the head turns.

Nature thus set up a double joint for the movements of the head, one between the atlas and axis for rotatory movements, another between the atlas and skull for nodding and side-to-side movements. And all these she increased by giving flexibility to the whole length of the neck. Makers of modern telescopes have imitated the method Nature invented when fixing the human head to the spine. Their instruments are mounted with a double joint—one for movements in a horizontal plane, the other for movements in a vertical plane. We thus see that the young engineer, as well as the student of medicine, can learn something from the construction of the human body.

In low forms of vertebrate animals like the fish and frog, the head is joined directly to the body, there being no neck.

No matter what part of the human body we examine, we shall find that its mechanical work is performed by means of bony levers. Having seen how the head is moved as a lever of the first order, we are now to choose a part which will show us the plan on which levers of the second order work, and there are many reasons why we should select the foot. It is a part which we are all familiar with; every day we can see it at rest and in action. The foot, as we have already noted, serves as a lever in walking. It is a bent or arched lever (Fig. 6); when we stand on one foot, the whole weight of our body rests on the summit of the arch. We are thus going to deal with a lever of a complex kind.

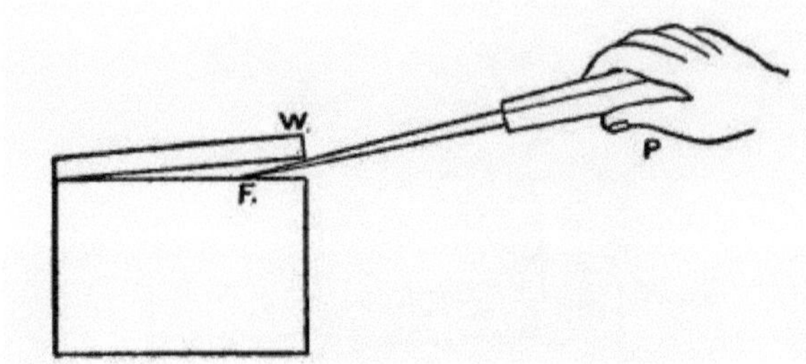

Fig. 5.—Showing a chisel used as a lever of the second order.

In using a chisel to pry open the lid of a box, we may use it as a lever either of the first or of the second order. We have already seen (Fig. 1) that, in using it as a lever of the first order, we pushed the handle downwards, while the bevelled end was raised, forcing open the lid. The edge of the box served as a rest or fulcrum for the chisel. If, however, after inserting the bevelled edge under the lid, we raise the handle instead of depressing it, we change the chisel into a lever of the second order. The lid is not now forced up on the bevelled edge, but is raised on the side of the chisel, some distance from the bevelled edge, which thus comes to represent the fulcrum. By using a chisel in this way, we reverse the positions of the weight and fulcrum and turn it into a lever of the second order. Suppose we push the side of the chisel—which is 10 inches long—under the lid to the extent of 1 inch, then the advantage we gain in power is as 1 to 10; we thereby increase our strength tenfold. If we push the chisel under the lid for half its length, then our advantage stands as 10 to 5; our strength is only doubled. If we push it still further for two-thirds of its length, then our gain in strength is only as 10 to 6.6; our power is increased by only one-third. Now this has an important bearing on the problem we are going to investigate, for the weight of our body falls on the foot, so that only about one-third of the lever—that part of it which is formed by the heel—projects behind the point on which the weight of the body rests. The strength of the muscles which act on the heel will be increased only by about one-third.

We have already seen that a double engine, made up of the *gastrocnemius* and *soleus*, is the power which is applied to the heel when we walk, and that the pad of the foot, lying across the sole in line with the ball of the great toe, serves as a fulcrum or rest. The weight of the body falls on the foot between the fulcrum in front and the power behind, as in a lever of the second order. We have explained why the power of the muscles of the calf is increased the more the weight of the body is shifted towards the toes, but it is also evident that the speed and the extent to which the body is lifted are diminished. If, however, the weight be shifted more towards the heel, the muscles of the calf, although losing in power, can lift their load more quickly and to a greater extent.

We must look closely at the foot lever if we are to understand it. It is arched or bent; the front pillar of the arch stretches from the summit or keystone, where the weight of the body is poised, to the pad of the foot or fulcrum (Fig. 6); the posterior pillar, projecting as the heel, extends from the summit to the point at which the muscular power is applied. A foot with a short anterior pillar and a long posterior pillar or heel is one designed for power, not speed. It is one which will serve a hill-climber well or a heavy, corpulent man. The opposite kind, one with a short heel and a long pillar in front, is well adapted for running and sprinting—for speed. Now, we do find among the various races of mankind that some have been given long heels, such as the dark-skinned natives of Africa and of Australia, while other races have been given relatively short, stumpy heels, of which sort the natives of Europe and of China may be cited as examples. With long heels less powerful muscular engines are required, and hence in dark races the calf of the leg is but ill developed, because the muscles which move the heel are small. We must admit, however, that the gait of dark-skinned races is usually easy and graceful. We Europeans, on the other hand, having short heels, need more powerful muscles to move them, and hence our calves are usually well developed, but our gait is apt to be jerky.

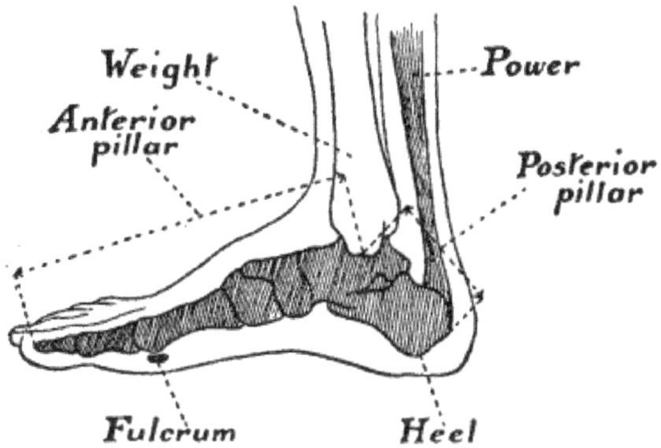

Fig. 6.—The bones forming the arch of the foot, seen from the inner side.

If we had the power to make our heels longer or shorter at will, we should be able, as is the case in a motor cycle, to alter our "speed-gear" according to the needs of the road. With a steep hill in front of us, we should adopt a long, slow, powerful heel; while going down an incline a short one would best suit our needs. With its four-change speed-gear a motor cycle seems better adapted for easy and economical travelling than the human machine. If, however, the human machine has no change of gear, it has one very marvellous mechanism—which we may call a *compensatory* mechanism, for want of a short, easy name. The more we walk, the more we go hill-climbing, the more powerful do the muscular engines of the heel become. It is quite different with the engine of a motor cycle; the more it is used, the more does it become worn out. It is because a muscular engine is living that it can respond to work by growing stronger and quicker.

I have no wish to extol the human machine unduly, nor to run down the motor cycle because of certain defects. There is one defect, however, which is inherent in all motor machines which man has invented, but from which the human machine is almost completely free. We can illustrate the defect best by comparing the movements of the heel with those of the crank-pin of an engine. One serves as the lever by which the gastrocnemius helps to propel the body; the other serves the same purpose in the propulsion of a motor cycle. On referring to Fig. 7, A, the reader will see that the piston-rod and the crank-pin are in a straight line; in such a position the engine is powerless to move the crank-pin until the flywheel is started, thus setting the crank-pin in motion. Once started, the leverage increases, until the crank-pin stands at right angles to the piston-rod—a point of maximum power which is reached when the piston is in the position shown in Fig. 7, B. Then the leverage decreases until the second dead centre is reached (Fig. 7, C); from that point the leverage is increased until the second maximum is reached (Fig. 7, D), whereafter it decreases until the arrival at the first position completes the cycle. Thus, in each revolution there are two points where all leverage or power is lost, points which are surmounted because of the momentum given by the flywheel. Clearly we should get most out of an engine if it could be kept working near the points of maximum leverage—with the lever as nearly as possible at right angles to the crank-pin.

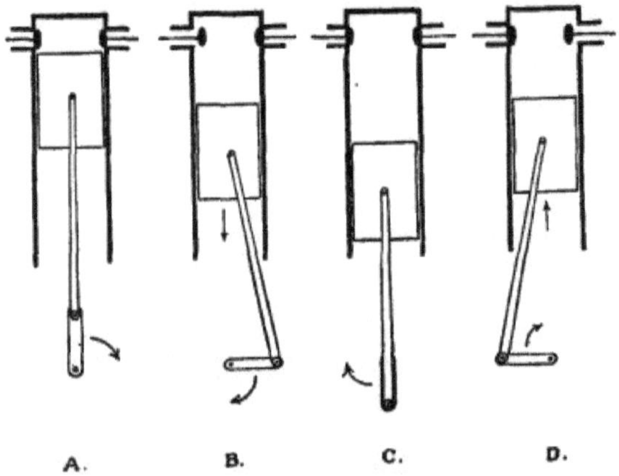

Fig. 7.—Showing the crank-pin of an engine at: A, First dead centre. B, First maximum leverage. C, Second dead centre. D, Second maximum leverage.

Now, we have seen that the tendon of Achilles is the piston cord, and the heel the crank-pin, of the muscular engine represented by the gastrocnemius and soleus. In the standing posture the heel slopes downwards and backwards, and is thus in a position, as regards its piston cord, considerably beyond the point of maximum leverage. As the heel is lifted by the muscles, it gradually becomes horizontal and at right angles to its tendon or piston cord. As the heel rises, then, it becomes a more effective lever; the muscles gain in power. The more the foot is arched, the more obliquely is the heel set and the greater is the strength needed to start it moving. Hence, races like the European and Mongolian, which have short as well as steeply set heels, need large calf muscles. It is at the end of the upward stroke that the heel becomes most effective as a lever, and it is just then that we most need power to propel our bodies in a forward direction. It will be noted that the heel, unlike the crank-pin of an engine, never reaches, never even approaches, that point of powerlessness known to engineers as a dead centre. Work is always performed within the limits of the most effective working radius of the lever. It is a law for all the levers of the body; they are set and moved in such a way as to avoid the occurrence of dead centres. Think what our condition would have been were this not so; why, we should require revolving fly-wheels set in all our joints!

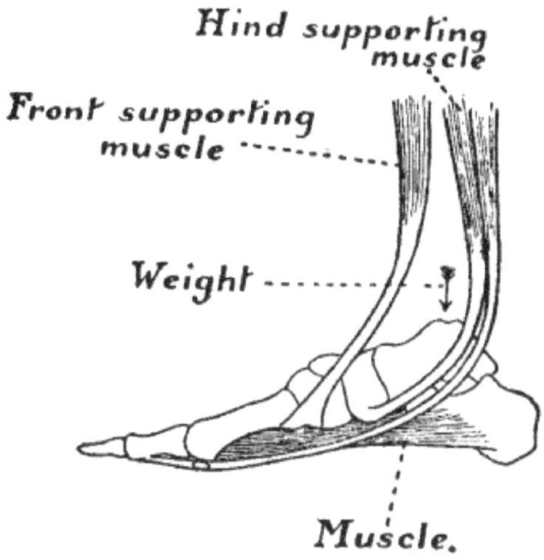

Fig. 8.—The arch of the foot from the inner side, showing some of the muscles which maintain it.

Another property is essential in a lever: it must be rigid; otherwise it will bend, and power will be lost. Now, if the foot were a rigid lever, there would be missing two of its most useful qualities. It could no longer act as a spring or buffer to the body, nor could it adapt its sole to the various kinds of surfaces on which we have to tread or stand. Nature, with her usual ingenuity, has succeeded in combining those opposing qualities—rigidity, suppleness, and elasticity or springiness—by resorting to her favorite device, the use of muscular engines. The arch is necessarily constructed of a number of bones which can move on each other to a certain extent, so that the foot may adapt itself to all kinds of roads and paths. It is true that the bones of the arch are loosely bound together by passive ties or ligaments, but as these cannot be lengthened or shortened at will, Nature had to fall back on the use of muscular engines for the maintenance of the foot as an arched lever. Some of these are shown in Fig. 8. The foot, then, is a lever of a very remarkable kind; all the time we stand or walk, its rigidity, its power to serve as a lever, has to be maintained by an elaborate battery of muscular engines all kept constantly at work. No wonder our feet and legs become tired when we have to stand a great deal. Some of these engines, the larger ones, are kept in the leg, but their tendons or piston cords descend below the ankle-joint to be fixed to various parts of the arch, and thus help to keep it up (Fig. 8). Within the sole of the foot has been placed an installation of seventeen small engines, all of them springing into action when we stand up, thus helping to maintain the foot as a rigid yet flexible lever.

We have already seen why our muscles are so easily exhausted when we stand stock-still; they then get no rest at all. Now, it sometimes happens in people who have to stand for long periods at a stretch that these muscular engines which maintain the arch are overtaxed; the arch of the foot gives way. The foot becomes flat and flexible, and can no longer serve as a lever. Many men and women thus become permanently crippled; they cannot step off their toes, but must shuffle along on the inner sides of their feet. But if the case of the overworked muscles which maintain the arch is hard in grown-up people, it is even harder in boys and girls who have to stand quite still for a long time, or who have to carry such burdens as are beyond their strength. When we are young, the bony levers and muscular engines of our feet have not only their daily work to do, but they have continually to effect those wonderful alterations which we call growth. Hence, the muscular engines of young people need special care; they must be given plenty of work to do, but that kind of active action which gives them alternate strokes of work and rest. Even the engine of a motor cycle has three strokes of play for one of work. Our engines, too, must have a liberal supply of the right kind of fuel. But even with all those precautions, we have to confess that the muscular engines of the foot do sometimes break down, and the leverage of the foot becomes threatened. Nor have we succeeded in finding out why they are so liable to break down in some boys and girls and not in others. Some day we shall discover this too.

We are now to look at another part of the human machine so that we may study a lever of the third order. The lever formed by the forearm and hand will suit our purpose very well. It is pivoted or jointed at the elbow; the elbow is its fulcrum (Fig. 9 B). At the opposite end of the lever, in the, upturned palm of the hand, we shall place a weight of 1 lb.

to represent the load to be moved. The power which we are to yoke to the lever is a strong muscular engine we have not mentioned before, called the *brachialis anticus*, or front brachial muscle. It lies in the upper arm, where it is fixed to the bone of that part—the humerus. It is attached to one of the bones of the forearm—the ulna—just beyond the elbow.

In the second order of lever, we have seen that the muscle worked on one end, while the weight rested on the lever somewhere between the muscular attachment and the fulcrum. In levers of the third order, the load is placed at the end of the lever, and the muscle is attached somewhere between the load and the fulcrum (Fig. 9 A). In the example we are considering, the brachial muscle is attached about half an inch beyond the fulcrum at the elbow, while the total length of the lever, measured from the elbow to the palm, is 12 inches. Now, it is very evident that the muscle or power being attached so close to the elbow, works under a great disadvantage as regards strength. It could lift a 24-lb. weight placed on the forearm directly over its attachment as easily as a single pound weight placed on the palm. But, then, there is this advantage: the 1-lb. weight placed in the hand moves with twenty-four times the speed of the 24-lb. weight situated near the elbow. What is lost in strength is gained in speed. Whenever Nature wishes to move a light load quickly, she employs levers of the third order.

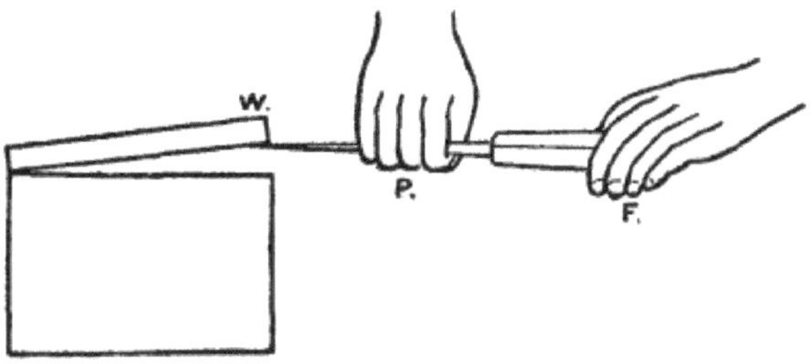

Fig. 9A.—A chisel used as a lever of the third order. W, weight; P, power; F, fulcrum.

We have often to move our forearm very quickly, sometimes to save our lives. The difference of one-hundredth of a second may mean life or death to us on the face of a cliff when we clutch at a branch or jutting rock to save a fall. The quickness of a blow we give or fend depends on the length of our reach. A long forearm and hand are ill adapted for lifting heavy burdens; strength is sacrificed if they are too long. Hence, we find that the laboring peoples of the world—Europeans and Mongolians—have usually short forearms and hands, while the peoples who live on such bounties as Nature may provide for them have relatively long forearms and hands.

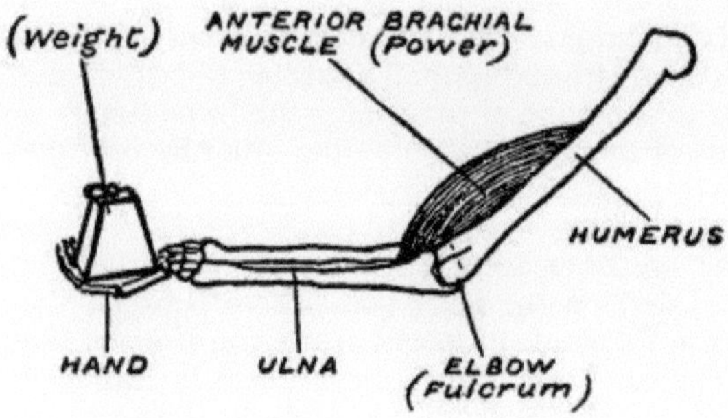

Fig. 9B.—The forearm and hand as a lever of the third order.

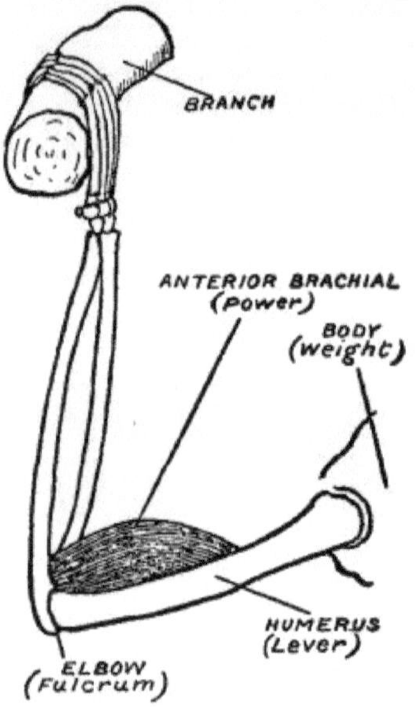

Fig. 10.—Showing the action of the brachialis anticus in the arm of an anthropoid ape.

Now, man differs from anthropoid apes, which are distant cousins of his, in having a forearm which is considerably shorter than the upper arm; whereas in anthropoid apes the forearm is much the longer. That fact surprises us at first, especially when we remember that anthropoids spend most of their lives amongst trees and use their arms much more than their legs in swinging the weight of their heavy bodies from branch to branch and from tree to tree. A long forearm and hand give them a long and quick reach, so that they can seize distant branches and swing themselves along safely and at a good pace. Our first thought is to suppose that a long forearm, being a weak lever, will be ill adapted for climbing. But when you look at Fig. 10, the explanation becomes plain. When a branch is seized by the hand, and the whole weight of the body is supported from it, the entire machinery of the arm changes its action. The forearm is no longer the lever which the brachial muscle moves (Fig. 10), but now becomes the base from which it acts. The part which was its piston cord now serves as its base of fixation, and what was its base of fixation to the humerus becomes its piston cord. The humerus has become a lever of the third order; its fulcrum is at the elbow; the weight of the body is attached to it at the shoulder and represents the load which has to be lifted. We also notice that the brachial muscle is attached a long way up the humerus, thus increasing its power very greatly, although the rate at which it helps in lifting the body is diminished. We can see, then, why the humerus is short and the forearm long in anthropoid apes; shortening the humerus makes it more powerful as a lever for lifting the body. That is why anthropoids are strong and agile tree-climbers. But then watch them use those long hands and forearms for the varied and precise movements we have to perform in our daily lives, and you will see how clumsy they are.

In the human machine the levers of the arm have been fashioned, not for climbing, but for work of another kind—the kind which brings us a livelihood. We must have perfect control over our hands; the longer the lever of the forearm is made, the more difficult does control of the hand become. Hence, in the human machine the forearm is made relatively short and the upper arm long.

We have just seen that the brachial muscle could at one time move the forearm and hand, but that when they are fixed it could then use the humerus as a lever and thereby lift the weight of the body. What should we think of a metal engine which could reverse its action so that it could act through its piston-rod at one time and through its cylinder at another? Yet that is what a great number of the muscular engines of the human machine do every day.

There is another little point, but an important one, which I must mention before this chapter is finished. I have spoken of the forearm and hand as if they formed a single solid lever. Of course that is not so; there are joints at the wrist where the hand can be moved on the forearm. But when a weight is placed in the hand, these joints became fixed by the action of muscles. The fixing muscles are placed in the forearm, both in front and behind, and are set in action the moment the hand is loaded. The wrist joint is fixed just in the same way as the joints of the foot are made rigid by muscles when it has to serve as a lever. Even when we take a pen in our hand and write, these engines which balance and fix the wrist have to be in action all the time. The steadiness of our writing depends on how delicately they are balanced. Like the muscles of the foot, the fixers of the wrist may become overworked and exhausted, as occasionally happens in men and women who do not hold their pens correctly and write for long spells day after day. The break-down which happens in them is called "writer's cramp," but it is a disaster of the same kind as that which overtakes the foot when its arch collapses, and its utility as a lever is lost.

THE EXPOSITION OF A MACHINE

THE MERGENTHALER LINOTYPE

Philip T. Dodge

The Mergenthaler Linotype machine appeared in crude form about 1886. This machine differs widely from all others in that it is adapted to produce the type-faces for each line properly justified on the edge of a solid slug or linotype.

These slugs, automatically produced and assembled by the machine, are used in the same manner as other type-forms, whether for direct printing or for electrotyping, and are remelted after use.

GENERAL ORGANIZATION

The general organization of the machine will first be described. After this the details will be more fully explained and attention plainly directed to the various parts which require special consideration.

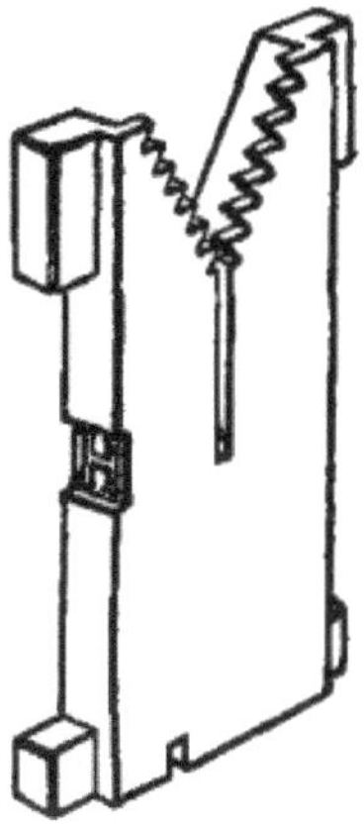

Fig. 1.

The machine contains, as the vital element, about sixteen hundred matrices, such as are shown in Fig. 1, each consisting of a small brass plate having in one edge the female character or matrix proper, and in the upper end a series of teeth, used as hereinafter explained for distributing the matrices after use to their proper places in the magazine of the machine. There are in the machine a number of matrices for each letter and also matrices representing special characters, and spaces or quadrats of different thicknesses for use in table-work. There is a series of finger keys representing the various characters and spaces, and the machine is so organized that on manipulating the keys it selects the matrices in the order in which their characters are to appear in print, and assembles them in a line, with wedge-shaped spaces or justifiers between the words. The series of matrices thus assembled in line forms a line matrix, or, in other words, a line of female dies adapted to mold or form a line of raised type on a slug cast against the matrices. After the matrix line is composed, it is automatically transferred to the face of a slotted mold into which molten type-metal is delivered to form a slug or linotype against the matrices. This done, the matrices are returned to the magazine and distributed, to be again composed in new relations for succeeding lines.

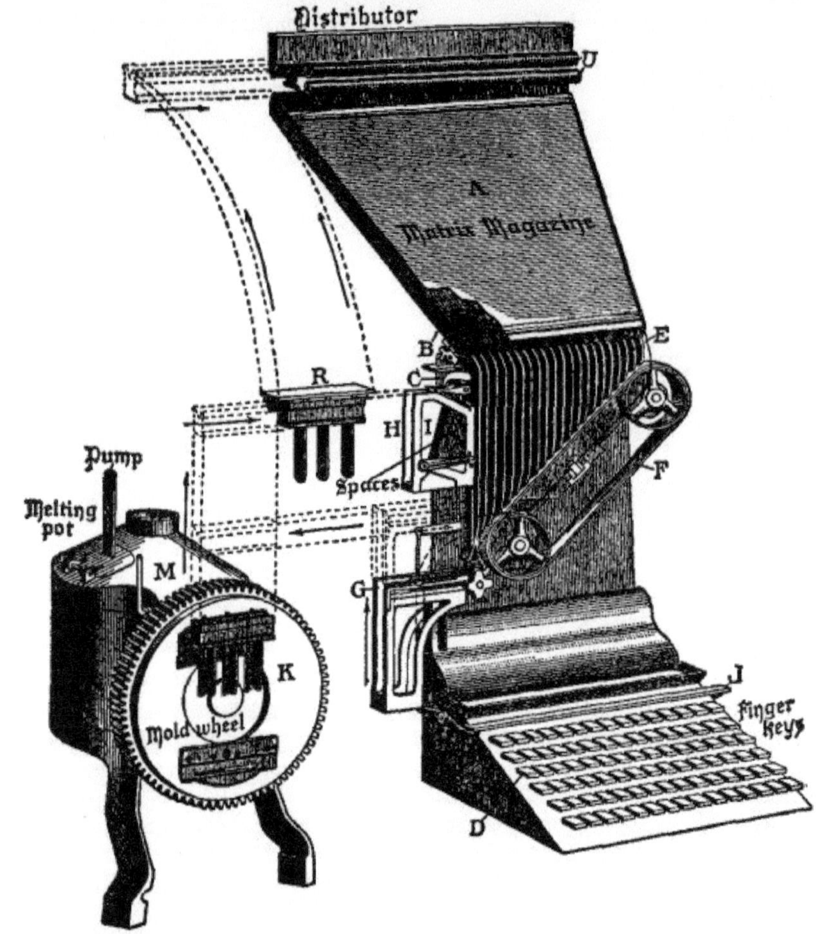

Fig. 2.

Fig. 2 illustrates the general organization of the machine.

A represents an inclined channelled magazine in which the matrices are stored. Each channel has at the lower end an escapement *B* to release the matrices one at a time. Each of these escapements is connected by a rod *C* and intermediate devices to one of the finger-keys in the keyboard *D*. These keys represent the various characters as in a typewriter. The keys are depressed in the order in which the characters and spaces are to appear, and the matricies, released successively from the lower end of the magazine, descend between the guides *E* to the surface of an inclined travelling belt *F*, by which they are carried downward and delivered successively into a channel in the upper part of the assembling elevator *G*, in which they are advanced by a star-shaped wheel, seen at the right.

The wedge-shaped spaces or justifiers *I* are held in a magazine *H*, from which they are delivered at proper intervals by finger-key *J* in the keyboard, so that they may pass downward and assume their proper positions in the line of matrices.

When the composition of the line is completed, the assembling elevator *G* is raised and the line is transferred, as indicated by dotted lines, first to the left and then downward to the casting position in front of the slotted mold seated in and extending through the vertical wheel *K*, as shown in Figs. 2 and 3. The line of matrices is pressed against and closes the front of the mold, the characters on the matrices standing directly opposite the slot in the mold, as shown. The back of the mold communicates with and is closed by the mouth of a melting-pot *M*, containing a supply of molten metal and heated by a Bunsen burner underneath. Within the pot is a vertical pump-plunger which acts at the proper time to drive the molten metal through the perforated mouth of the pot into the mold and into all the characters in the matrices. The metal, solidifying, forms a slug or linotype bearing on its edge, in relief, type-characters produced from the matrices. The matrices and the pot are immediately separated from the mold, and the mold wheel rotates until the slug contained in the mold is presented in front of an ejector blade, where the slug is ejected from the mold through a pair of knives, which trim the sides to the required size, into the receiving galley, as shown in Fig. 4.

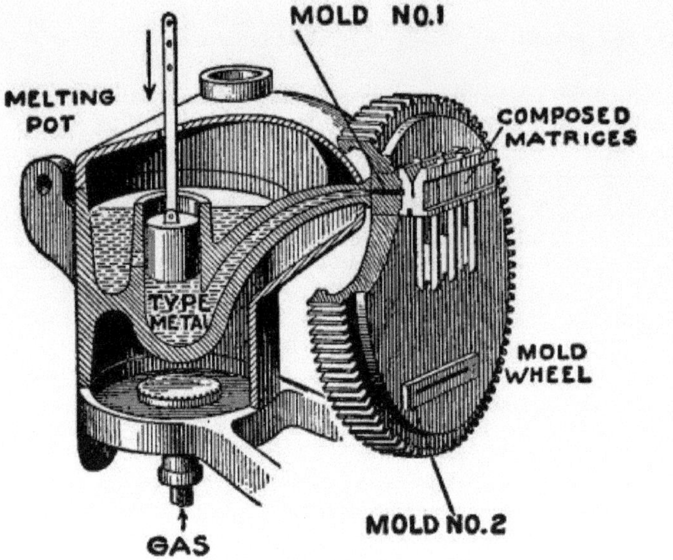

Fig. 3.

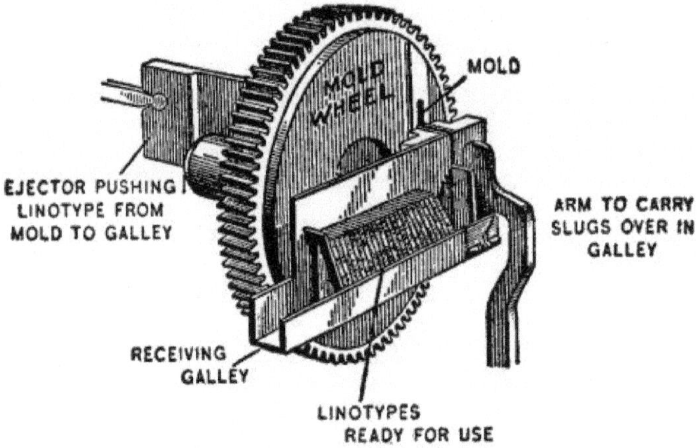

Fig. 4.

After the line of matrices and spaces has served its purpose, it is raised from the casting position and moved to the right, as shown by the dotted lines and arrows in Fig. 2. The teeth in the upper ends of the matrices are engaged with a toothed bar *R*, known as the second elevator. This elevator swings upward, as shown by dotted lines, carrying the matrices to the level of the upper end of the magazine, and leaving the spaces or justifiers behind to be transferred to their magazine *H*.

The distributing mechanism consists essentially of a fixed bar *T*, lying in a horizontal position above the upper end of the magazine, and having along its lower edge, as shown in Fig. 2, horizontal teeth to engage the teeth in the upper end of the matrices and hold them in suspension. The teeth of the matrix for each letter differ in number or arrangement, or both, from the teeth of matrices bearing other letters, and the teeth on the lower edge of the distributor bar are correspondingly varied in arrangement at different points in the length of the bar. (See Fig. 2.)

The matrices are moved forward into engagement with the distributor bar and also into engagement with the threads of horizontal screws *U*, which are extended parallel with the distributor bar and constantly rotated so that they cause the matrices to travel one after another along the distributor and over the mouths of the channels in the magazines. Each matrix is held in suspension until it arrives over its proper channel, where for the first time its teeth bear such relation to those of the bar that it is released and permitted to fall into the magazine.

The speed of the machine, which is commonly from four to five thousand ems per hour, but which has reached ten thousand and upward in competitive trials, is due to the fact that the matrices pursue a circulatory course, leaving the magazine at the lower end, passing thence to the line and to the casting mechanism, and finally returning to the top of the magazine. This permits the composition of one line, the casting of another, and the distribution of a third to proceed simultaneously.

ASSEMBLING AND KEYBOARD MECHANISMS

The matrices pass through the magazine by gravity. Their release is effected by mechanisms shown in Figs. 5 and 6, which are vertical sections through the magazine, the keyboard, and intermediate connections. Under each channel of the magazine, there is an escapement *B*, consisting of a small lever rocking at its centre on a horizontal pivot, and carrying at its opposite ends two dogs or pawls *b*, *b*, which are projected up alternately into the magazine by the motion of the lever. The key-rod *C*, suspended from the rear end of the escapement *B*, tends to hold the lower pawl *b* in an elevated position, as shown in Fig. 5, so that it engages under the upper ear of the foremost matrix to prevent its escape.

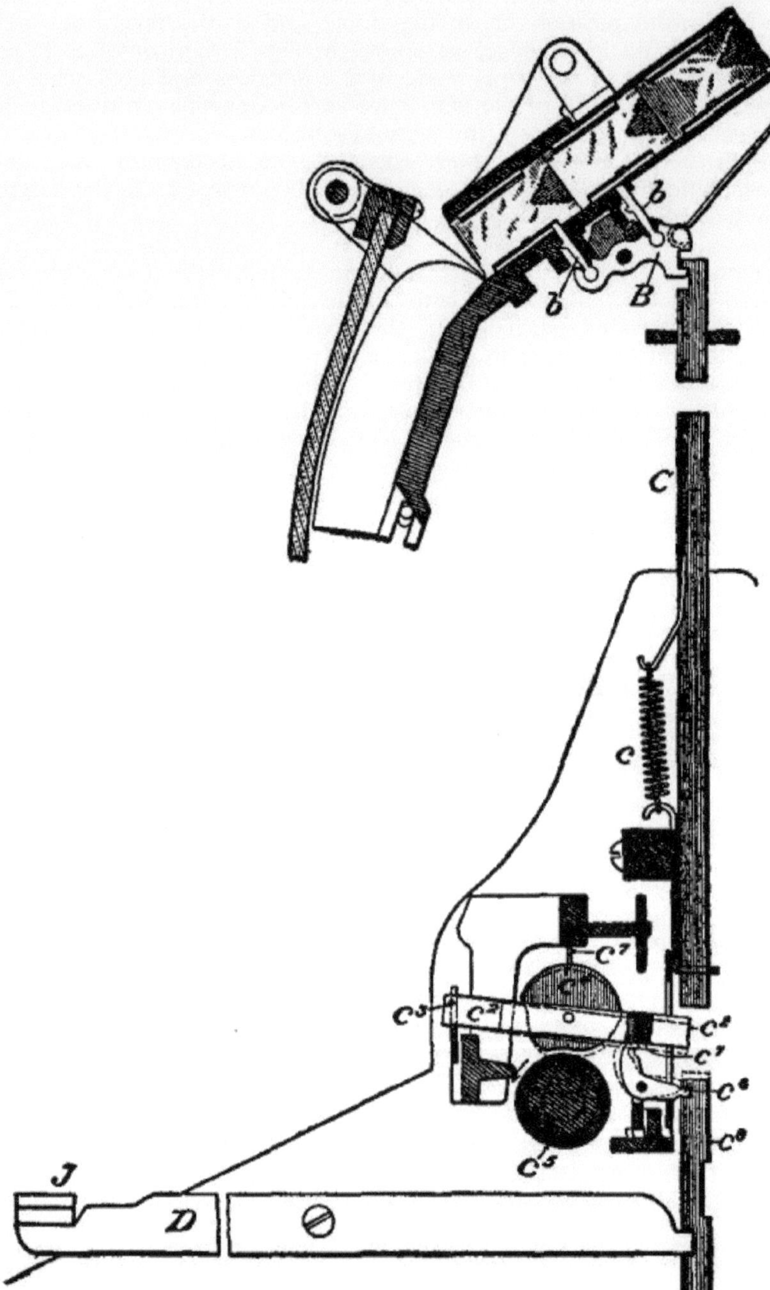

b'
b
b
B
C
c
C7
C5
C2
C4
C2
C7
C6
C8
C5
J
D

Fig. 5.

When the escapement *B* is rocked, it withdraws the lower pawl *b*, as shown in Fig. 6, at the same time raising the upper pawl, so that it engages and momentarily arrests the next matrix. As soon as the first matrix has escaped, the escapement resumes its original position, the upper pawl falling, while the lower one rises so as to hold the second matrix, which assumes the position previously occupied by the one released.

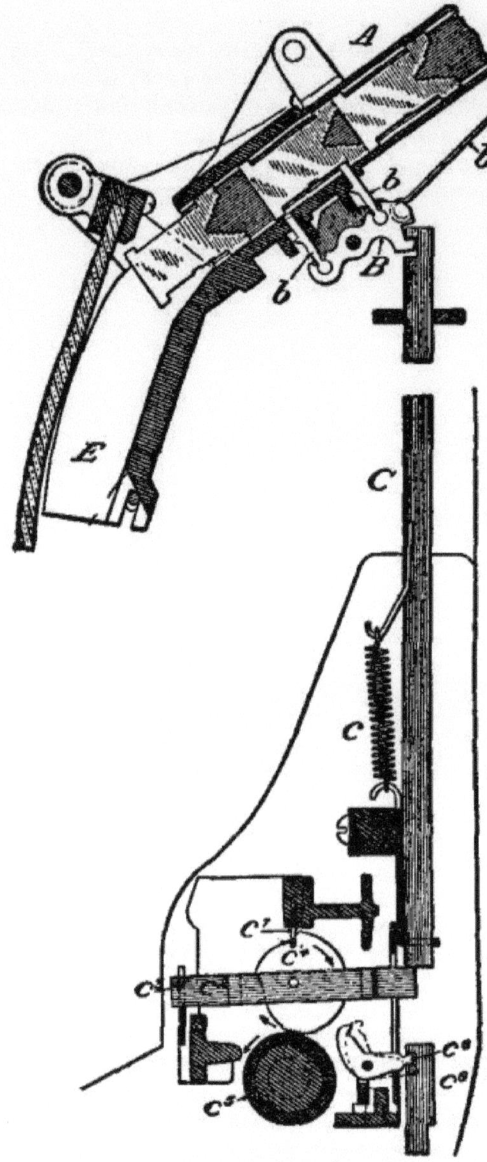

Fig. 6.

Thus it is that the alternate rising and falling of the two escapement pawls permits the matrices to escape one at a time. It is evident that the escapements could be operated directly by rods connected with the finger-keys, but this direct connection is objectionable because of the labor required on the part of the operator, and the danger that the keys may not be fully depressed.

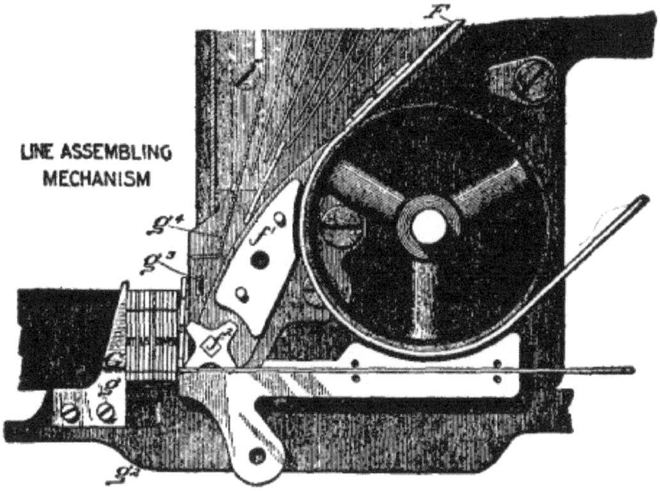

Fig. 7.

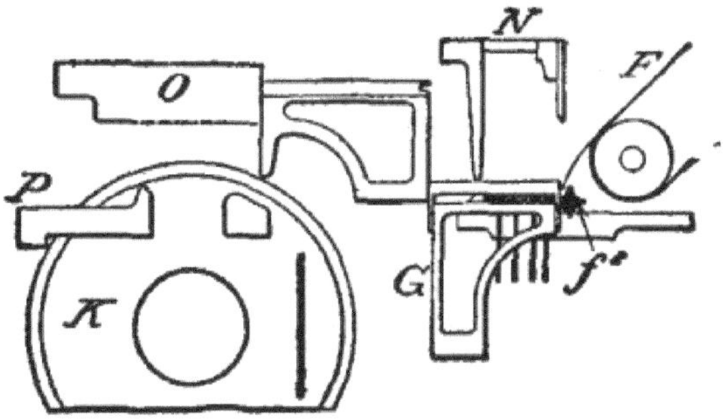

Fig. 8.

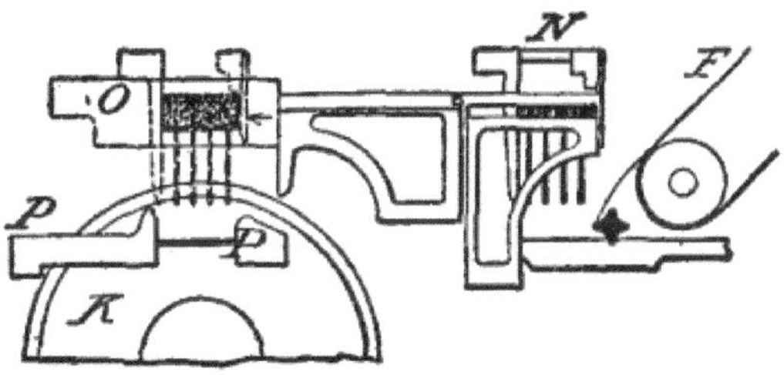

Fig. 9.

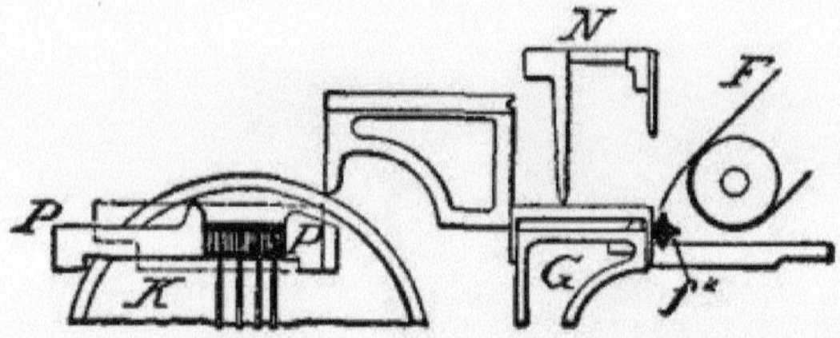

Fig. 10. Moreover, it is essential that the escapements should act individually with moderate speed to the end that the matrices may be properly engaged and disengaged by the pawls. For these reasons, and to secure easy and uniform action of the parts, the mechanism shown in Figs. 5 and 6 is introduced between the finger-keys and escapements. The vertical rods *C*, which actuate the escapements, are guided in the main frame, and each is urged downward by a spring *c*. Each rod *C* terminates directly over one end of a rising and falling yoke-bar c^2, turning on a pivot c^3 at the opposite end. Each of the yokes c^2 is slotted vertically to admit an eccentric c^4 turning on a pivot therein. A constantly rotating rubber-covered roll c^5 is extended across the entire keyboard beneath the cams, which stand normally as shown in Fig. 5, out of contact with the roll. When the parts are in this position, the cam-yoke is sustained at its free end by the yoke-trigger c^8, and a cross-bar in the cam engages a vertical pin c^7 on the frame, whereby the cam is prevented from falling on to the roller, as it has a tendency to do. Each of the yoke-triggers c^6 is connected with a vertical bar c^8, which is in turn connected to the rear end of a finger-key lever *D*. The parts stand normally at rest in the position shown in Fig. 5, the roll c^5 turning freely under the cam without effect upon it.

When the finger-key is depressed, it raises the bar c^8, which in turn trips the yoke-trigger c^6 from under the cam-yoke c^2, permitting the latter to fall, thereby lowering the cam c^4 into peripheral engagement with the rubber roll, at the same time disengaging the cam from the stop-pin c^7. The roll, engaging frictionally with the cam, causes the latter to turn on its centre in the direction indicated by the arrow in Fig. 6.

Owing to the eccentric shape of the cam, its rotation while resting on the roller causes it to lift the yoke c^2 above its original position, so that it acts upon the escapement rod *C*, lifting it and causing it to reverse the position of the escapement *B*, to release the matrix, as plainly seen in Fig. 6.

While this is taking place, the yoke-trigger c^6 resumes its first position, as shown in dotted lines in Fig. 6, so that as the rotating cam lowers the yoke, it is again supported in its first position, the cam at the same time turning forward by momentum out of engagement with the roll until arrested in its original position by the pin c^7.

It will be observed that the parts between each key lever and escapement operate independently of the others, so that a number of cams may be in engagement with the rollers at one time, and a number of escapements at different stages of their action at one time.

The matrices falling from the magazine descend through the front channels and are received on the inclined belt *F*, on which they are carried over and guided on the upper rounding surface of the assembler entrance-block f^1, by which they are guided downward in front of the star-wheel f^2, which pushes them forward one after another.

The spaces or justifiers *I*, released from their magazine *H*, as heretofore described, descend into the assembler *G* in front of the star-wheel in the same manner as the matrices.

The line in course of composition is sustained at its front end by a yielding finger or resistant *g*, secured to a horizontal assembler slide g^2, the purpose of these parts being to hold the line together in compact form.

As the matrices approach the line, their upper ends are carried over a spring g^3, projecting through the assembler face-plate from the rear, as shown in Fig. 7, its purpose being to hold the matrices forward and prevent them from falling back in such a manner that succeeding matrices and spaces or justifiers will pass improperly ahead of them. The descending matrices also pass beneath a long depending spring g^4, which should be so adjusted as barely to permit the passage of the thickest matrix.

After the composition of the line is completed in the assembling elevator *G*, as shown in Fig. 8, the elevator is raised as shown in Fig. 9, so as to present the line between the depending fingers of the transfer-carriage *N*, which then moves to the left to the position shown by dotted lines in Fig. 9, thereby bringing the line into the first elevator *O*, which then descends, carrying the line of matrices downwards, as shown in Fig. 10, to its position in front of the mold and between the confining jaws *P*, *P*, mounted in the main frame, which determine the length of the line.

Figs. 11 and 12 show the casting mechanism in vertical section from front to rear. When the first elevator *O* lowers the line, as just described, the mold and the pot *M* stand in their rearward positions, as shown in Fig. 11.

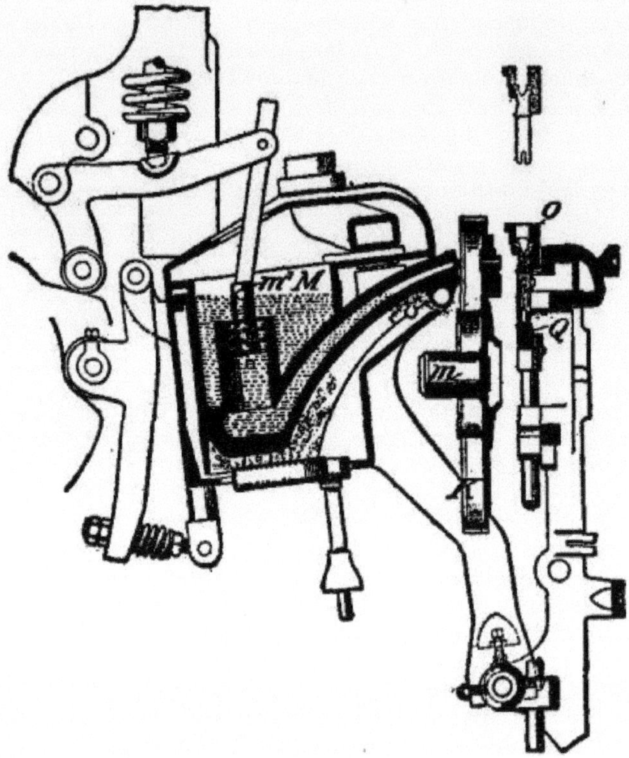

Fig. 11.

The mold-carrying wheel is sustained by a horizontal slide, and as soon as the matrix line is lowered to the casting position, a cam at the rear pushes the slide and mold wheel forward until the front face of the mold is closed tightly against the rear face of the matrix line, as shown in Fig. 12.

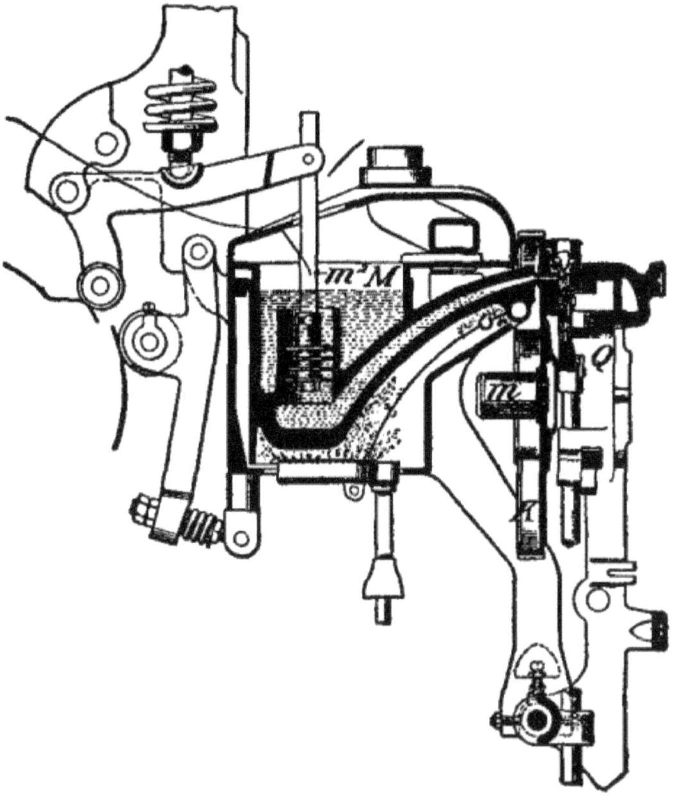

Fig. 12.

While this is taking place, the pot, having its supporting legs mounted on a horizontal shaft, swings forward until its mouth is closed tightly against the back of the mold, as shown in Fig. 12. While the parts are in this position, the justifying bar *Q* is driven up and pushes the spaces or justifiers upward through the line of matrices until the line is expanded or elongated to fill completely the gap between jaws *P*, *P*.

In order to secure exact alignment of the matrices vertically and horizontally, the bar *Q* acts repeatedly on the spaces, and the line is slightly unlocked endwise and relocked. This is done that the matrices may be temporarily released to facilitate the accurate adjustment demanded. While the justified line is locked fast between the jaws, the elevator, and the mold, the plunger m^2 in the pot descends and drives the molten metal before it through the spout or mouth of the pot into the mold, which is filled under pressure, so that a solid slug is produced against the matrices. The pot then retreats, and its mouth breaks away from the back of the slug in the mold, while, at the same time, the mold retreats to draw the type-characters on the contained slug out of the matrices. The mold wheel now revolves, carrying the rear edge of the slug past a stationary trimming-knife, not shown, and around to the position in front of the ejector, as previously described and shown in Fig. 4, whereupon the ejector advances and drives the slug between two side trimming-knives into the galley at the front.

DISTRIBUTION

After the casting action the first elevator *O* rises and carries the matrix line above the original or composing level, as shown in Fig. 13. The line is then drawn horizontally to the right until the teeth of the matrices engage the toothed elevator bar *R*, which swings upward with the matrices, thus separating the matrices from the spaces or justifiers *I*, which remain suspended in the frame, so that they may be pushed to the right, as indicated by the arrow, into their magazine.

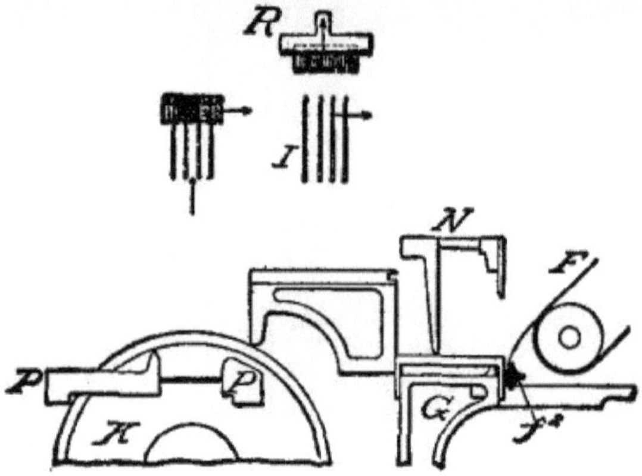

Fig. 13.

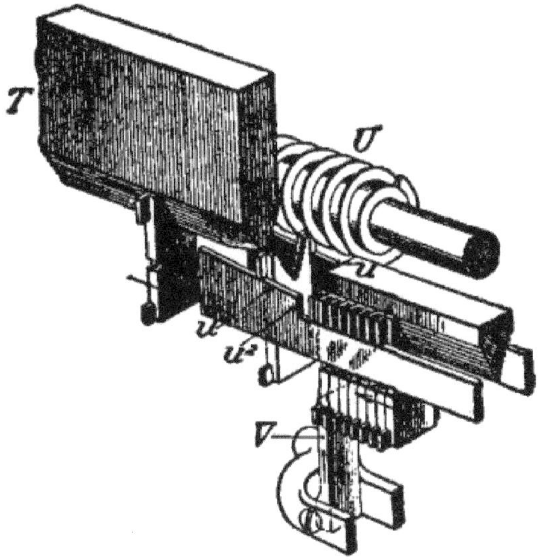

Fig. 14.

When the line of matrices is raised to the distributor, it is necessary that the matrices shall be separated and presented one at a time to the distributor bar, between the threads of the horizontal carrier-screws. This is accomplished as shown in Figs. 14 and 15. A horizontal pusher or line-shifter *S* carries the line of matrices forward from the elevator bar *R* into the so-called distributor box, containing at its opposite sides two rails *u*, having near their forward ends shoulders u^2, against which the forward matrix abuts so as to prevent further advance of the line, which is urged constantly forward by the follower or line-shifter *S*. A vertically reciprocating lifting finger *V* has its upper end shouldered to engage beneath the foremost matrix, so as to push it upward until its upper ears are lifted above the detaining shoulder u^2, so that they may ride forward on the upwardly inclined inner ends of the rails, as shown in Fig. 14. The matrices thus lifted are engaged by the screws and carried forward, and, as they move forward, they are gradually raised by the rails until the teeth finally engage themselves on the distributor bar *T*, from which they are suspended as they are carried forward, over the mouth of the magazine, until they fall into their respective channels, as shown in Fig. 15.

The distributor box also contains on opposite sides shorter rails, u^4, adapted to engage the lower ends of the matrices, to hold them in position as they are lifted. The lifting finger *V* is mounted on a horizontal pivot in one end of an elbow lever mounted on pivot v^2 and actuated by a cam on the end of one of the carrier-screws, as shown in Figs. 2 and 15.

TRIMMING-KNIVES

In practice there is occasionally found a slight irregularity in the thickness of slugs, and thin fins are sometimes cast around the forward edges. For the purpose of reducing them to a uniform thickness, they are driven on their way to the galley between two vertical knives, as shown in Figs. 4 and 16. The inner knife is stationary, but the outer knife is adjustable in order that it may accommodate slugs of different thicknesses. This adjustment is made by the knife being seated at its outer edge against a supporting bar or wedge, having at opposite ends two inclined surfaces seated against supporting screws in the knife-block. A lever engages a pin on the wedge for the purpose of moving it endwise; when moving in one direction, it forces the knife inward toward the stationary knife, and when moved in the other direction, it forces it to retreat under the influence of a spring seated in the block. The wedge is provided with a series of teeth engaged by a spring-actuated pin or dog, whereby the wedge and the knife are stopped in proper positions to insure the exact space required between the two knives.

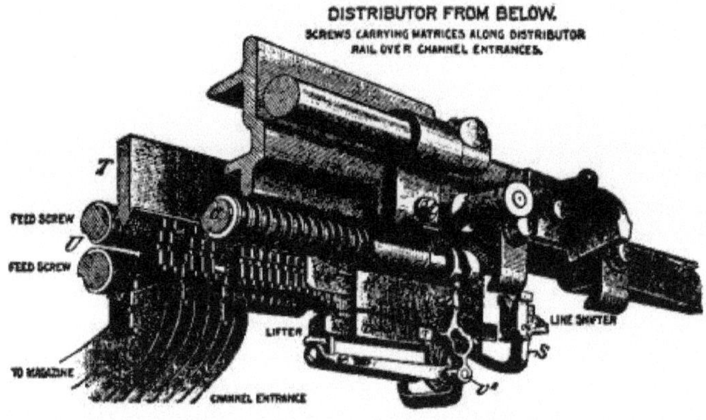

Fig. 15.

The back knife, secured to the frame for trimming the base of the slug as it is carried past by the revolving wheel, should be kept moderately sharp and adjusted so as to fit closely against the back of the passing mold. Particular attention should be paid to this feature. The edge of the knife must bear uniformly across the face of the mold.

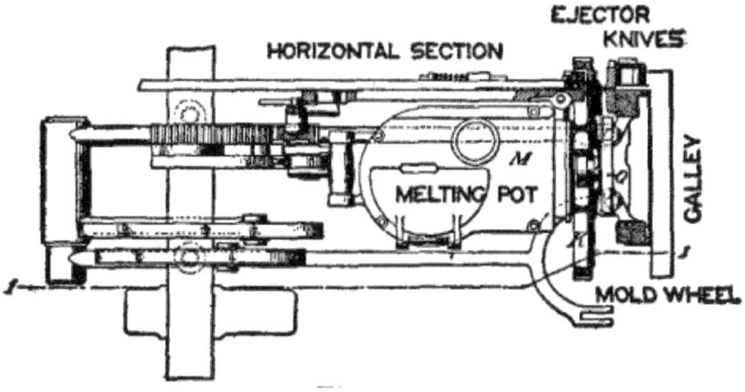

Fig. 16.

The front knives, between which the slug is ejected, should not be made too sharp. After being sharpened, the thin edge can be advantageously removed by the use of a thin oilstone applied against the side face; that is, against the face past which the slug is carried.

The stationary or left-hand knife should be so adjusted as to align exactly with the inner side of the mold. Under proper conditions this knife does not trim the side face of the slug, but acts only to remove any slight fins or projections at the front edge.

The right-hand knife, adjustable by means of a wedge and lever, should stand exactly parallel with the stationary knife. It trims the side of the slug on which the ribs are formed, and it serves to bring the slug to the

THE EXPOSITION OF A PROCESS IN NATURE

THE PEA WEEVIL

Jean Henri Fabre

Peas are held in high esteem by mankind. From remote ages man has endeavored, by careful culture, to produce larger, tenderer, and sweeter varieties. Of an adaptable character, under careful treatment the plant has evolved in a docile fashion, and has ended by giving us what the ambition of the gardener desired. To-day we have gone far beyond the yield of the Varrons and Columelles, and further still beyond the original pea; from the wild seeds confided to the soil by the first man who thought to scratch up the surface of the earth, perhaps with the half-jaw of a cave-bear, whose powerful canine tooth would serve him as a ploughshare!

Where is it, this original pea, in the world of spontaneous vegetation? Our own country has nothing resembling it. Is it to be found elsewhere? On this point botany is silent, or replies only with vague probabilities.

We find the same ignorance elsewhere on the subject of the majority of our alimentary vegetables. Whence comes wheat, the blessed grain which gives us bread? No one knows. You will not find it here, except in the care of man; nor will you find it abroad. In the East, the birthplace of agriculture, no botanist has ever encountered the sacred ear growing of itself on unbroken soil.

Barley, oats, and rye, the turnip and the beet, the beetroot, the carrot, the pumpkin, and so many other vegetable products, leave us in the same perplexity; their point of departure is unknown to us, or at most suspected behind the impenetrable cloud of the centuries. Nature delivered them to us in the full vigor of the thing untamed, when their value as food was indifferent, as to-day she offers us the sloe, the bullace, the blackberry, the crab; she gave them to us in the state of imperfect sketches, for us to fill out and complete; it was for our skill and our labor patiently to induce the nourishing pulp which was the earliest form of capital, whose interest is always increasing in the primordial bank of the tiller of the soil.

As storehouses of food the cereal and the vegetable are, for the greater part, the work of man. The fundamental species, a poor resource in their original state, we borrowed as they were from the natural treasury of the vegetable world; the perfected race, rich in alimentary materials, is the result of our art.

If wheat, peas, and all the rest are indispensable to us, our care, by a just return, is absolutely necessary to them. Such as our needs have made them, incapable of resistance in the bitter struggle for survival, these vegetables, left to themselves without culture, would rapidly disappear, despite the numerical abundance of their seeds, as the foolish sheep would disappear were there no more sheep-folds.

They are our work, but not always our exclusive property. Wherever food is amassed, the consumers collect from the four corners of the sky; they invite themselves to the feast of abundance, and the richer the food the greater their numbers. Man, who alone is capable of inducing agrarian abundance, is by that very fact the giver of an immense banquet at which legions of feasters take their place. By creating more juicy and more generous fruits, he calls to his enclosures, despite himself, thousands and thousands of hungry creatures, against whose appetites his prohibitions are helpless. The more he produces, the larger is the tribute demanded of him. Wholesale agriculture and vegetable abundance favor our rival, the insect.

This is the immanent law. Nature, with an equal zeal, offers her mighty breast to all her nurslings alike; to those who live by the goods of others no less than to the producers. For us, who plough, sow, and reap, and weary ourselves with labor, she ripens the wheat; she ripens it also for the little Calender-beetle, which, although exempted from the labor of the fields, enters our granaries none the less, and there, with its pointed beak, nibbles our wheat, grain by grain, to the husk.

For us, who dig, weed, and water, bent with fatigue and burned by the sun, she swells the pods of the pea; she swells them also for the weevil, which does no gardener's work, yet takes its share of the harvest at its own hour, when the earth is joyful with the new life of spring.

Let us follow the manoeuvres of this insect which takes its tithe of the green pea. I, a benevolent rate-payer, will allow it to take its dues; it is precisely to benefit it that I have sown a few rows of the beloved plant in a corner of my garden. Without other invitation on my part than this modest expenditure of seed-peas, it arrives punctually during the month of May. It has learned that this stony soil, rebellious at the culture of the kitchen-gardener, is bearing peas for the first time. In all haste therefore it has hurried, an agent of the entomological revenue system, to demand its dues.

Whence does it come? It is impossible to say precisely. It has come from some shelter, somewhere, in which it has passed the winter in a state of torpor. The plane-tree, which sheds its rind during the heats of the summer, furnishes an excellent refuge for homeless insects under its partly detached sheets of bark.

I have often found our weevil in such a winter refuge. Sheltered under the dead covering of the plane, or otherwise protected while the winter lasts, it awakens from its torpor at the first touch of a kindly sun. The almanac of the instincts has aroused it; it knows as well as the gardener when the pea-vines are in flower, and seeks its favorite plant, journeying thither from every side, running with quick, short steps, or nimbly flying.

A small head, a fine snout, a costume of ashen grey sprinkled with brown, flattened wing-covers, a dumpy, compact body, with two large black dots on the rear segment—such is the summary portrait of my visitor. The middle of May approaches, and with it the van of the invasion.

They settle on the flowers, which are not unlike white-winged butterflies. I see them at the base of the blossom or inside the cavity of the "keel" of the flower, but the majority explore the petals and take possession of them. The time for laying the eggs has not yet arrived. The morning is mild; the sun is warm without being oppressive. It is the moment of nuptial flights; the time of rejoicing in the splendor of the sunshine. Everywhere are creatures rejoicing to be alive. Couples come together, part, and re-form. When towards noon the heat becomes too great, the weevils retire into the shadow, taking refuge singly in the folds of the flowers whose secret corners they know so well. To-morrow will be another day of festival, and the next day also, until the pods, emerging from the shelter of the "keel" of the flower, are plainly visible, enlarging from day to day.

A few gravid females, more pressed for time than the others, confide their eggs to the growing pod, flat and meager as it issues from its floral sheath. These hastily laid batches of eggs, expelled perhaps by the exigencies of an ovary incapable of further delay, seem to me in serious danger; for the seed in which the grub must establish itself is as yet no more than a tender speck of green, without firmness and without any farinaceous tissue. No larva could possibly find sufficient nourishment there, unless it waited for the pea to mature.

But is the grub capable of fasting for any length of time when once hatched? It is doubtful. The little I have seen tells me that the newborn grub must establish itself in the midst of its food as quickly as possible, and that it perishes unless it can do so. I am therefore of opinion that such eggs as are deposited in immature pods are lost. However, the race will hardly suffer by such a loss, so fertile is the little beetle. We shall see directly how prodigal the female is of her eggs, the majority of which are destined to perish.

The important part of the maternal task is completed by the end of May, when the shells are swollen by the expanding peas, which have reached their final growth, or are but little short of it. I was anxious to see the female Bruchus at work in her quality of Curculionid, as our classification declares her. The other weevils are Rhyncophora, beaked insects, armed with a drill with which to prepare the hole in which the egg is laid. The Bruchus possesses only a short snout or muzzle, excellently adapted for eating soft tissues, but valueless as a drill.

The method of installing the family is consequently absolutely different. There are no industrious preparations as with the Balinidae, the Larinidae, and the Rhynchitides. Not being equipped with a long oviscapt, the mother sows her eggs in the open, with no protection against the heat of the sun and the variations of temperature. Nothing could be simpler, and nothing more perilous to the eggs, in the absence of special characteristics which, would enable them to resist the alternate trials of heat and cold, moisture and drought.

In the caressing sunlight of ten o'clock in the morning, the mother runs up and down the chosen pod, first on one side, then on the other, with a jerky, capricious, unmethodical gait. She repeatedly extrudes a short oviduct, which oscillates right and left as though to graze the skin of the pod. An egg follows, which is abandoned as soon as laid.

A hasty touch of the oviduct, first here, then there, on the green skin of the pea-pod, and that is all. The egg is left there, unprotected, in the full sunlight. No choice of position is made such as might assist the grub when it seeks to penetrate its larder. Some eggs are laid on the swellings created by the peas beneath; others in the barren valleys which separate them. The first are close to the peas, the second at some distance from them. In short, the eggs of the Bruchus are laid at random, as though on the wing.

We observe a still more serious vice: the number of eggs is out of all proportion to the number of peas in the pod. Let us note at the outset that each grub requires one pea; it is the necessary ration, and is largely sufficient for one larva, but is not enough for several, nor even for two. One pea to each grub, neither more nor less, is the unchangeable rule.

We should expect to find signs of a procreative economy which would impel the female to take into account the number of peas contained in the pod which she has just explored; we might expect her to set a numerical limit on her eggs in conformity with that of the peas available. But no such limit is observed. The rule of one pea to one grub is always contradicted by the multiplicity of consumers.

My observations are unanimous on this point. The number of eggs deposited on one pod always exceeds the number of peas available, and often to a scandalous degree. However meager the contents of the pod, there is a superabundance of consumers. Dividing the sum of the eggs upon such or such a pod by that of the peas contained therein, I find there are five to eight claimants for each pea; I have found ten, and there is no reason why this prodigality should not go still further. Many are called, but few are chosen! What is to become of all these supernumeraries, perforce excluded from the banquet for want of space?

The eggs are of a fairly bright amber yellow, cylindrical in form, smooth, and rounded at the ends. Their length is at most a twenty-fifth of an inch. Each is affixed to the pod by means of a slight network of threads of coagulated albumen. Neither wind nor rain can loosen their hold.

The mother not infrequently emits them two at a time, one above the other; not infrequently, also, the uppermost of the two eggs hatches before the other, while the latter fades and perishes. What was lacking to this egg, that it should fail to produce a grub? Perhaps a bath of sunlight; the incubating heat of which the outer egg has robbed it. Whether on account of the fact that it is shadowed by the other egg, or for other reasons, the elder of the eggs in a group of two rarely follows the normal course, but perishes on the pod, dead without having lived.

There are exceptions to this premature end; sometimes the two eggs develop equally well; but such cases are exceptional, so that the Bruchid family would be reduced to about half its dimensions if the binary system were the rule. To the detriment of our peas and to the advantage of the beetle, the eggs are commonly laid one by one and in isolation.

A recent emergence is shown by a little sinuous ribbon-like mark, pale or whitish, where the skin of the pod is raised and withered, which starts from the egg and is the work of the newborn larva; a sub-epidermic tunnel along which the grub works its way, while seeking a point from which it can escape into a pea. This point once attained, the larva, which is scarcely a twenty-fifth of an inch in length, and is white with a black head, perforates the envelope and plunges into the capacious hollow of the pod.

It has reached the peas and crawls upon the nearest. I have observed it with the magnifier. Having explored the green globe, its new world, it begins to sink a well perpendicularly into the sphere. I have often seen it halfway in, wriggling its tail in the effort to work the quicker. In a short time the grub disappears and is at home. The point of entry, minute, but always easily recognizable by its brown coloration on the pale green background of the pea, has no fixed location; it may be at almost any point on the surface of the pea, but an exception is usually made of the lower half; that is, the hemisphere whose pole is formed by the supporting stem.

It is precisely in this portion that the germ is found, which will not be eaten by the larva, and will remain capable of developing into a plant, in spite of the large aperture made by the emergence of the adult insect. Why is this particular portion left untouched? What are the motives that safeguard the germ?

It goes without saying that the Bruchus is not considering the gardener. The pea is meant for it and for no one else. In refusing the few bites that would lead to the death of the seed, it has no intention of limiting its destruction. It abstains from other motives.

Let us remark that the peas touch laterally, and are pressed one against the other, so that the grub, when searching for a point of attack, cannot circulate at will. Let us also note that the lower pole expands into the umbilical excrescence, which is less easy of perforation than those parts protected by the skin alone. It is even possible that the umbilicum, whose organization differs from that of the rest of the pea, contains a peculiar sap that is distasteful to the little grub.

Such, doubtless, is the reason why the peas exploited by the Bruchus are still able to germinate. They are damaged, but not dead, because the invasion was conducted from the free hemisphere, a portion less vulnerable and more easy of access. Moreover, as the pea in its entirety is too large for a single grub to consume, the consumption is limited to the portion preferred by the consumer, and this portion is not the essential portion of the pea.

With other conditions, with very much smaller or very much larger seeds, we shall observe very different results. If too small, the germ will perish, gnawed like the rest by the insufficiently provisioned inmate; if too large, the abundance of food will permit of several inmates. Exploited in the absence of the pea, the cultivated vetch and the broad bean afford us an excellent example; the smaller seed, of which all but the skin is devoured, is left incapable of germination; but the large bean, even though it may have held a number of grubs, is still capable of sprouting.

Knowing that the pod always exhibits a number of eggs greatly in excess of the enclosed peas, and that each pea is the exclusive property of one grub, we naturally ask what becomes of the superfluous grubs. Do they perish outside when the more precocious have one by one taken their places in their vegetable larder? or do they succumb to the intolerant teeth of the first occupants? Neither explanation is correct. Let us relate the facts.

On all old peas—they are at this stage dry—from which the adult Bruchus has emerged, leaving a large round hole of exit, the magnifying-glass will show a variable number of fine reddish punctuations, perforated in the centre. What are these spots, of which I count five, six, and even more on a single pea? It is impossible to be mistaken: they are the points of entry of as many grubs. Several grubs have entered the pea, but of the whole group only one has survived, fattened, and attained the adult age. And the others? We shall see.

At the end of May, and in June, the period of egg-laying, let us inspect the still green and tender peas. Nearly all the peas invaded show us the multiple perforations already observed on the dry peas abandoned by the weevils. Does this actually mean that there are several grubs in the pea? Yes. Skin the peas in question, separate the cotyledons, and break them up as may be necessary. We shall discover several grubs, extremely youthful, curled up comma-wise, fat and lively, each in a little round niche in the body of the pea.

Peace and welfare seem to reign in the little community. There is no quarrelling, no jealousy between neighbors. The feast has commenced; food is abundant, and the feasters are separated one from another by the walls of uneaten substance. With this isolation in separate cells no conflicts need be feared; no sudden bite of the mandibles, whether intentional or accidental. All the occupants enjoy the same rights of property, the same appetite, and the same strength. How does this communal feast terminate?

Having first opened them, I place a number of peas which are found to be well peopled in a glass test-tube. I open others daily. In this way I keep myself informed as to the progress of the various larvae. At first nothing noteworthy is to be seen. Isolated in its narrow chamber, each grub nibbles the substance around it, peacefully and parsimoniously. It is still very small; a mere speck of food is a feast; but the contents of one pea will not suffice the whole number to the end. Famine is ahead, and all but one must perish.

Soon, indeed, the aspect of things is entirely changed. One of the grubs—that which occupies the central position in the pea—begins to grow more quickly than the others. Scarcely has it surpassed the others in size when the latter cease to eat, and no longer attempt to burrow forwards. They lie motionless and resigned; they die that gentle death which comes to unconscious lives. Henceforth the entire pea belongs to the sole survivor. Now what has happened that these lives around the privileged one should be thus annihilated? In default of a satisfactory reply, I will propose a suggestion.

In the centre of the pea, less ripened than the rest of the seed by the chemistry of the sun, may there not be a softer pulp, of a quality better adapted to the infantile digestion of the grub? There, perhaps, being nourished by tenderer, sweeter, and perhaps, more tasty tissues, the stomach becomes more vigorous, until it is fit to undertake less easily digested food. A nursling is fed on milk before proceeding to bread and broth. May not the central portion of the pea be the feeding-bottle of the Bruchid?

With equal rights, fired by an equal ambition, all the occupants of the pea bore their way towards the delicious morsel. The journey is laborious, and the grubs must rest frequently in their provisional niches. They rest; while resting they frugally gnaw the riper tissues surrounding them; they gnaw rather to open a way than to fill their stomachs.

Finally one of the excavators, favored by the direction taken, attains the central portion. It establishes itself there, and all is over; the others have only to die. How are they warned that the place is taken? Do they hear their brother gnawing at the walls of his lodging? can they feel the vibration set up by his nibbling mandibles? Something of the kind must happen, for from that moment they make no attempt to burrow further. Without struggling against the fortunate winner, without seeking to dislodge him, those which are beaten in the race give themselves up to death. I admire this candid resignation on the part of the departed.

Another condition—that of space—is also present as a factor. The pea weevil is the largest of our Bruchidae. When it attains the adult stage, it requires a certain amplitude of lodging, which the other weevils do not require in the same degree. A pea provides it with a sufficiently spacious cell; nevertheless, the cohabitation of two in one pea would be impossible; there would be no room, even were the two to put up with a certain discomfort. Hence the necessity of an inevitable decimation, which will suppress all the competitors save one.

Now the superior volume of the broad bean, which is almost as much beloved by the weevil as the pea, can lodge a considerable community, and the solitary can live as a cenobite. Without encroaching on the domain of their neighbors, five or six or more can find room in the one bean.

Moreover, each grub can find its infant diet; that is, that layer which, remote from the surface, hardens only gradually and remains full of sap until a comparatively late period. This inner layer represents the crumb of a loaf, the rest of the bean being the crust.

In a pea, a sphere of much less capacity, it occupies the central portion; a limited point at which the grub develops, and lacking which it perishes; but in the bean it lines the wide adjoining faces of the two flattened cotyledons. No matter where the point of attack is made, the grub has only to bore straight down when it quickly reaches the softer tissues. What is the result? I have counted the eggs adhering to a bean-pod and the beans included in the pod, and comparing the two figures I find that there is plenty of room for the whole family at the rate of five or six dwellers in each bean. No superfluous larvae perish of hunger when barely issued from the egg; all have their share of the ample provision; all live and prosper. The abundance of food balances the prodigal fertility of the mother.

If the Bruchus were always to adopt the broad bean for the establishment of her family, I could well understand the exuberant allowance of eggs to one pod; a rich foodstuff easily obtained evokes a large batch of eggs. But the case of the pea perplexes me. By what aberration does the mother abandon her children to starvation on this totally insufficient vegetable? Why so many grubs to each pea when one pea is sufficient only for one grub?

Matters are not so arranged in the general balance-sheet of life. A certain foresight seems to rule over the ovary so that the number of mouths is in proportion to the abundance or scarcity of the food consumed. The Scarabaeus, the Sphex, the Necrophorus, and other insects which prepare and preserve alimentary provision for their families, are all of a narrowly limited fertility, because the balls of dung, the dead or paralyzed insects, or the buried corpses of animals on which their offspring are nourished are provided only at the cost of laborious efforts.

The ordinary bluebottle, on the contrary, which lays her eggs upon butcher's meat or carrion, lays them in enormous batches. Trusting in the inexhaustible riches represented by the corpse, she is prodigal of offspring, and takes no account of numbers. In other cases the provision is acquired by audacious brigandage, which exposes the newly born offspring to a thousand mortal accidents. In such cases the mother balances the chances of destruction by an exaggerated flux of eggs. Such is the case with the Meloides, which, stealing the goods of others under conditions of the greatest peril, are accordingly endowed with a prodigious fertility.

The Bruchus knows neither the fatigues of the laborious, obliged to limit the size of her family, nor the misfortunes of the parasite, obliged to produce an exaggerated number of offspring. Without painful search, entirely at her ease, merely moving in the sunshine over her favorite plant, she can insure a sufficient provision for each of her offspring; she can do so, yet is foolish enough to over-populate the pod of the pea; a nursery insufficiently provided, in which the great majority will perish of starvation. This ineptitude is a thing I cannot understand; it clashes too completely with the habitual foresight of the maternal instinct.

I am inclined to believe that the pea is not the original food plant of the Bruchus. The original plant must rather have been the bean, one seed of which is capable of supporting a dozen or more larvae. With the larger cotyledon the crying disproportion between the number of eggs and the available provision disappears.

Moreover, it is indubitable that the bean is of earlier date than the pea. Its exceptional size and its agreeable flavor would certainly have attracted the attention of man from the remotest periods. The bean is a ready-made mouthful, and would be of the greatest value to the hungry tribe. Primitive man would at an early date have sown it beside his wattled hut. Coming from Central Asia by long stages, their wagons drawn by shaggy oxen and rolling on the circular discs cut from the trunks of trees, the early immigrants would have brought to our virgin land, first the bean, then the pea, and finally the cereal, that best of safeguards against famine. They taught us the care of herds, and the use of bronze, the material of the first metal implement. Thus the dawn of civilization arose over France. With the bean did those ancient teachers also involuntarily bring us the insect which to-day disputes it with us? It is doubtful; the Bruchidae seem to be indigenous. At all events, I find them levying tribute from various indigenous plants, wild vegetables which have never tempted the appetite of man. They abound in particular upon the great forest vetch (*Lathyrus latifolius*), with its magnificent heads of flowers and long handsome pods. The seeds are not large, being indeed smaller than the garden pea; but, eaten to the very skin, as they invariably are, each is sufficient to the needs of its grub.

We must not fail to note their number. I have counted more than twenty in a single pod, a number unknown in the case of the pea, even in the most prolific varieties. Consequently this superb vetch is in general able to nourish without much loss the family confided to its pod.

Where the forest vetch is lacking, the Bruchus, none the less, bestows its habitual prodigality of eggs upon another vegetable of similar flavor, but incapable of nourishing all the grubs: for example, the travelling vetch (*Vicia peregrina*) or the cultivated vetch (*Vicia saliva*). The number of eggs remains high even upon insufficient pods, because the original food-plant offered a copious provision, both in the multiplicity and the size of the seeds. If the Bruchus is really a stranger, let us regard the bean as the original food-plant; if indigenous, the large vetch.

Sometime in the remote past we received the pea, growing it at first in the prehistoric vegetable garden which already supplied the bean. It was found a better article of diet than the broad bean, which to-day, after such good service, is comparatively neglected. The weevil was of the same opinion as man, and without entirely forgetting the bean and the vetch it established the greater part of its tribe upon the pea, which from century to century was more widely cultivated. To-day we have to share our peas; the Bruchidae take what they need, and bestow their leavings on us.

This prosperity of the insect which is the offspring of the abundance and equality of our garden products is from another point of view equivalent to decadence. For the weevil, as for ourselves, progress in matters of food and drink is not always beneficial. The race would profit better if it remained frugal. On the bean and the vetch the Bruchus founded colonies in which the infant mortality was low. There was room for all. On the pea-vine, delicious though its fruits may be, the greater part of its offspring die of starvation. The rations are few, and the hungry mouths are multitudinous.

We will linger over this problem no longer. Let us observe the grub which has now become the sole tenant of the pea by the death of its brothers. It has had no part in their death; chance has favored it, that is all. In the centre of the pea, a wealthy solitude, it performs the duty of a grub, the sole duty of eating. It nibbles the walls enclosing it, enlarging its lodgment, which is always entirely filled by its corpulent body. It is well shaped, fat, and shining with health. If I disturb it, it turns gently in its niche and sways its head. This is its manner of complaining of my importunities. Let us leave it in peace.

It profits so greatly and so swiftly by its position that by the time the dog-days have come it is already preparing for its approaching liberation. The adult is not sufficiently well equipped to open for itself a way out through the pea, which is now completely hardened. The larva knows of this future helplessness, and with consummate art provides for its release. With its powerful mandibles it bores a channel of exit, exactly round, with extremely clean-cut sides. The most skilful ivory-carver could do no better.

To prepare the door of exit in advance is not enough; the grub must also provide for the tranquillity essential to the delicate processes of nymphosis. An intruder might enter by the open door and injure the helpless nymph. This passage must therefore remain closed. But how?

As the grub bores the passage of exit, it consumes the farinaceous matter without leaving a crumb. Having come to the skin of the pea, it stops short. This membrane, semi-translucid, is the door to the chamber of metamorphosis, its protection against the evil intentions of external creatures.

It is also the only obstacle which the adult will encounter at the moment of exit. To lessen the difficulty of opening it, the grub takes the precaution of gnawing at the inner side of the skin, all round the circumference, so as to make a line of least resistance. The perfect insect will only have to heave with its shoulder and strike a few blows with its head in order to raise the circular door and knock it off like the lid of a box. The passage of exit shows through the diaphanous skin of the pea as a large circular spot, which is darkened by the obscurity of the interior. What passes behind it is invisible, hidden as, it is behind a sort of ground-glass window.

A pretty invention, this little closed porthole, this barricade against the invader, this trap-door raised by a push when the time has come for the hermit to enter the world. Shall we credit it to the Bruchus? Did the ingenious insect conceive the undertaking? Did it think out a plan and work out a scheme of its own devising? This would be no small triumph for the brain of a weevil. Before coming to a conclusion, let us try an experiment.

I deprive certain occupied peas of their skin, and I dry them with abnormal rapidity, placing them in glass test-tubes. The grubs prosper as well as in the intact peas. At the proper time the preparations for emergence are made.

If the grub acts on its own inspiration, if it ceases to prolong its boring directly it recognizes that the outer coating, auscultated from time to time, is sufficiently thin, what will it do under the conditions of the present test? Feeling itself at the requisite distance from the surface, it will stop boring; it will respect the outer layer of the bare pea, and will thus obtain the indispensable protecting screen.

Nothing of the kind occurs. In every case the passage is completely excavated; the entrance gapes wide open, as large and as carefully executed as though the skin of the pea were in its place. Reasons of security have failed to modify the usual method of work. This open lodging has no defence against the enemy; but the grub exhibits no anxiety on this score.

Neither is it thinking of the outer enemy when it bores down to the skin when the pea is intact, and then stops short. It suddenly stops because the innutritious skin is not to its taste. We ourselves remove the parchment-like skins from a mess of pease-pudding, as from a culinary point of view they are so much waste matter. The larva of the Bruchus, like ourselves, dislikes the skin of the pea. It stops short at the horny covering, simply because it is checked by an uneatable substance. From this aversion a little miracle arises; but the insect has no sense of logic; it is passively obedient to the superior logic of facts. It obeys its instinct, as unconscious of its act as is a crystal when it assembles, in exquisite order, its battalions of atoms.

Sooner or later during the month of August we see a shadowy circle form on each inhabited pea; but only one on each seed. These circles of shadow mark the doors of exit. Most of them open in September. The lid, as though cut out with a punch, detaches itself cleanly and falls to the ground, leaving the orifice free. The Bruchus emerges, freshly clad, in its final form.

The weather is delightful. Flowers are abundant, awakened by the summer showers; and the weevils visit them in the lovely autumn weather. Then, when the cold sets in, they take up their winter quarters in any suitable retreat. Others, still numerous, are less hasty in quitting the native seed. They remain within during the whole winter, sheltered behind the trap-door, which they take care not to touch. The door of the cell will not open on its hinges, or, to be exact, will not yield along the line of least resistance, until the warm days return. Then the late arrivals will leave their shelter and rejoin the more impatient, and both will be ready for work when the pea-vines are in flower.

To take a general view of the instincts in their inexhaustible variety is, for the observer, the great attraction of the entomological world, for nowhere do we gain a clearer sight of the wonderful way in which the processes of life are ordered. Thus regarded, entomology is not, I know, to the taste of everybody; the simple creature absorbed in the doings and habits of insects is held in low esteem. To the terrible utilitarian, a bushel of peas preserved from the weevil is of more importance than a volume of observations which bring no immediate profit.

Yet who has told you, O man of little faith, that what is useless to-day will not be useful to-morrow? If we learn the customs of insects or animals, we shall understand better how to protect our goods. Do not despise disinterested knowledge, or you may rue the day. It is by the accumulation of ideas, whether immediately applicable or otherwise, that humanity has done, and will continue to do, better to-day than yesterday, and better to-morrow than to-day. If we live on peas and beans, which we dispute with the weevil, we also live by knowledge, that mighty kneading-trough in which the bread of progress is mixed and leavened. Knowledge is well worth a few beans.

Among other things, knowledge tells us: "The seedsman need not go to the expense of waging war upon the weevil. When the peas arrive in the granary, the harm is already done; it is irreparable, but not transmissible. The untouched peas have nothing to fear from the neighborhood of those which have been attacked, however long the mixture is left. From the latter the weevils will issue when their time has come; they will fly away from the storehouse if escape is possible; if not, they will perish without in any way attacking the sound peas. No eggs, no new generation will ever be seen upon or within the dried peas in the storehouse; there the adult weevil can work no further mischief."

The Bruchus is not a sedentary inhabitant of granaries: it requires the open air, the sun, the liberty of the fields. Frugal in everything, it absolutely disdains the hard tissues of the vegetable; its tiny mouth is content with a few honeyed mouthfuls, enjoyed upon the flowers. The larvae, on the other hand, require the tender tissues of the green pea growing in the pod. For these reasons the granary knows no final multiplication on the part of the despoiler.

The origin of the evil is in the kitchen-garden. It is there that we ought to keep a watch on the misdeeds of the Bruchus, were it not for the fact that we are nearly always weaponless when it comes to fighting an insect. Indestructible by reason of its numbers, its small size, and its cunning, the little creature laughs at the anger of man. The gardener curses it, but the weevil is not disturbed; it imperturbably continues its trade of levying tribute. Happily we have assistants more patient and more clear-sighted than ourselves.

During the first week of August, when the mature Bruchus begins to emerge, I notice a little Chalcidian, the protector of our peas. In my rearing-cages it issues under my eyes in abundance from the peas infested by the grub of the weevil. The female has a reddish head and thorax; the abdomen is black, with a long augur-like oviscapt. The male, a little smaller, is black. Both sexes have reddish claws and thread-like antennae.

In order to escape from the pea, the slayer of the weevil makes an opening in the centre of the circular trap-door which the grub of the weevil prepared in view of its future deliverance. The slain has prepared the way for the slayer. After this detail the rest may be divined.

When the preliminaries to the metamorphosis are completed, when the passage of escape is bored and furnished with its lid of superficial membrane, the female Chalcidian arrives in a busy mood. She inspects the peas, still on the vine, and enclosed in their pods; she auscultates them with her antennae; she discovers, hidden under the general envelope, the weak points in the epidermic covering of the peas. Then, applying her oviscapt, she thrusts it through the side of the pod and perforates the circular trap-door. However far withdrawn into the centre of the pea, the Bruchus, whether larvae or nymph, is reached by the long oviduct. It receives an egg in its tender flesh, and the thing is done. Without possibility of defence, since it is by now a somnolent grub or a helpless pupa, the embryo weevil is eaten until nothing but skin remains. What a pity that we cannot at will assist the multiplication of this eager exterminator! Alas! our assistants have got us in a vicious circle, for if we wished to obtain the help of any great number of Chalcidians we should be obliged in the first place to breed a multiplicity of Bruchidae.

THE EXPOSITION OF A MANUFACTURING PROCESS

MODERN PAPER-MAKING

J.W. Butler Paper Company

Though the steady march of progress and invention has given to the modern paper-maker marvelous machines by which the output is increased a thousandfold over that of the old, slow methods, he still has many of the same difficulties to overcome that confronted his predecessor. While the use of wood pulp has greatly changed the conditions as regards the cheaper grades of this staple, the ragman is to-day almost as important to the manufacturer of the higher grades as he was one hundred years ago when the saving of rags was inculcated as a domestic virtue and a patriotic duty. Methods have changed, but the material remains the same. In a complete modern mill making writing and other high-grade papers, the process begins with unsightly rags as the material from which to form the white sheets that are to receive upon their spotless polished surface the thoughts of philosophers and statesmen, the tender messages of affection, the counsels and admonitions of ministers, the decisions of grave and learned judges, and all the

Wisdom of things, mysterious, divine, thatIllustriously doth on paper shine,

as was duly set forth in rhyme by the *Boston News Letter* in 1769. "The bell cart will go through Boston about the end of next month," it announced, and appealed to the inhabitants of that modern seat of learning and philosophy to save their rags for the occasion, and thus encourage the industry.

The rags do not come to the mammoth factories of to-day in bell carts, but by the carload in huge bales gathered from all sections of this great Republic, as well as from lands beyond the eastern and western oceans. The square, compact, steam-compressed bundles are carried by elevators well up toward the top of the building, where they await the knife of the "opener." When they have been opened, the "feeder" throws the contents by armfuls into the "thrasher." The novice or layman, ignorant of the state in which rags come to the mill, will find their condition a most unpleasant surprise, especially disagreeable to his olfactory nerves. Yet the unsavory revelation comes with more force a little farther on, in the "assorting-room." The "thrasher" is a great cylindrical receptacle, revolving rapidly, which is supplied with long wooden beaters or arms passing through a wooden cylinder and driven by power. When the rags have been tossed in, there ensues a great pounding and thrashing, and the dust is carried off in suction air-tubes, while the whipped rags are discharged and carried to the "sorting" and "shredding" room. Here the rags are assorted as to size, condition, and the presence of buttons, hooks and eyes, or other material that must be removed. Then those that need further attention are passed on to the "shredders," these as well as the "sorters" being women. The "shredders" stand along a narrow counter; in front of each one there is fastened a long scythe-blade with its back toward the operator and its point extending upward, the shank being firmly fixed to the table or operating board. Here buttons, hard seams, and all similar intruders are disposed of, and the larger pieces of rags are cut into numerous small ones on the scythe-blades. The rags thus prepared are tossed by the women into receptacles in the tables. The work in this room is the most disagreeable and unwholesome in the entire process of manufacture, and this despite the fact that these rags, too, have been thrashed, and freed from an amount of dust and dirt beyond belief.

While one is watching the operations carried on here, it is impossible to repress the wish that rags might be bought otherwise than by the pound, for, unfortunately, filth, dust, and dirt weigh, and to wash rags only reduces the weight. While this is a true reflection of the condition in the average mill, it is pleasant to know, however, there are others of the higher class that are decided exceptions as far as dust and dirt are concerned. Such are the mills making high-grade ledger and bond papers, as well as the mill manufacturing the paper that is used for the printing of our "greenbacks," to which further reference will be made later. In these exceptional mills everything is neat and perfectly clean, all the stock used being new and fresh from the cotton or linen mills, or from factories producing cloth goods, like shirt and corset factories, and others of the same sort. The sorting and shredding room is always large and light, with windows on all sides, and well ventilated, offering a decided contrast in many respects to the less cleanly mills first referred to where the women must wear bonnets or hoods for the protection of the hair. In either case the process is certainly an improvement over the old plan of leaving the rags to decay in a cellar to expedite the removal of the glutinous matter from them.

From the "sorting" and "shredding" room the rags are conveyed to the "cutter," where they are cut and chopped by revolving knives, leaving them in small pieces and much freer from dust and grit. Various ingenious devices are employed for removing metal and other hard and injurious matter, magnetic brushes serving this purpose in some mills. When the "cutter" has finished its work, the still very dirty rags go for a further cleansing to the "devil," or "whipper," a hollow cone with spikes projecting within, against which work the spikes of a drum, dashing the rags about at great speed. Human lives are often freed of their baser elements and restored to purity and beauty through the chastening influences of tribulation or adversity; in like manner the "whipper" carries the rags forward a step in the process of purification that is necessary before they can be brought to their highest usefulness. But the cleansing process, which is only a preparation for what is to follow, does not end with the "whipper," which has served merely to loosen, not to dislodge, a great deal of dust and dirt. The final operation in the preliminary cleaning is performed by the "duster" proper, which is a conical revolving sieve. As the mass of rags is tossed and shaken about, the loosened dust is carried away by the suction of the air, which draws the dust particles into tubes furnished with suction fans. In most modern mills the rags are carried forward from the "duster" on an endless belt, and a careful watch is kept upon them as they emerge to detect the presence of unchopped pieces, buttons, or other foreign substances. The journey of the rags over this endless belt or conveyor terminates in a receiving-room, in the floor of which there are several openings, and immediately below these the mouths of the "digesters," which are in a room beneath. The "digesters," as they are suggestively and appropriately termed, are huge revolving boilers, usually upright, which often have as great a diameter as eight feet, with a height of twenty-two feet and a digestive capacity of upward of five tons of rags each. The rags that are to be "cooked" are fed in to the "digesters" through the openings in the floor, and the great movable manhole plates are then put in place and closed, hermetically sealing the openings or mouths through which the boilers have been fed, these having first been charged with a mixed solution of lime and soda and with live hot steam in lieu of gastric juice as a digesting fluid and force. In some mills the boilers are placed in a horizontal position, while in others they are in the form of a large ball or globe, in either case being operated in the manner described; those of upright form, however, are most commonly in use. The rags are boiled under steam pressure of about forty pounds to the square inch, and the cooking is continued from twelve to fourteen hours.

It is here that the process of cleaning begins in earnest; and as the mass of rags is tumbled about in its scalding bath of steam-heated lime-water, or "milk of lime," the coloring and glutinous matters, as well as all other impurities, are loosened from the fibers, which are in the end so cleansed and purified as to come forth unstained and of virgin purity. Having been sufficiently boiled and digested, the mushy material, still looking dark and forbidding, is emptied onto the floor below or into receptacles placed directly beneath the boilers, where the color and dirt are allowed to drain off. The mass is then conveyed to the "washers," great tub-like receptacles, which are known as "Hollanders," from the fact that these rag engines were invented in Holland about the year 1750 A.D. They are oval-shaped tubs, about twenty feet long, nine feet wide, and three feet high, varying somewhat according to the conditions. Each tub is divided for two-thirds of its length by an upright partition, or "mid-feather," as it is called, which makes a narrow course around the vat. On one side of the partition, the tub is raised in a half-circle, close to which revolves an iron roll about three or four feet in diameter, and covered with knives; in the bottom of the tub, and directly under the revolving roll, is another set of knives called a "bed-plate," which is stationary, and against which the roll can be lowered. But let us not anticipate. When the emptyings from the boiler have been thrown into the "washer," a continuous stream of water is turned in at one end, the knife-roll having been adjusted so as to open up the rags as they are set in motion. These then begin a lively chase around the edge of the vat, through the race-course formed by the "mid-feather," and under the rag-opening knives, where the water is given a chance to wash out all impurities, then on up the incline over the "back-fall," so-called from the elevation in the tub. A cylinder of wire-cloth, partly immersed in the moving mass, holds back the now rapidly whitening fibers, while the dirty water escapes into buckets inside the wire-cloth drum, and is discharged into and through an escape-spout. The heavy particles of dirt settle into what is termed a "sand-trap" at the bottom of the tub.

As the water clears, the roll is lowered closer and closer to the bottom of the bed-plate, in order to open up the fibers more thoroughly for the free circulation of the water among them. When the several agencies of the "washer" have accomplished their purpose and the water runs clear and unsullied, a bleaching material is put into the mass, which in the course of from two to six hours becomes as white as milk. The dirty offscourings of all ragdom, first seen in the original bales, and gathered from the four corners of the globe, have endured many buffetings, many bruisings and tribulations, and having been washed come forth pure, sweet, and clean. From the washers the rags are precipitated through a trap into drainers, which are chambers made of stone and brick, with a false bottom through which the water is allowed to drain. This rag pulp, now called half stock, is kept in this receptacle until the water and liquor are thoroughly drained off, when it becomes a white and compact mass of fibers.

The rags should stand in the drainers for at least one week, though better results are obtained if they are left for a period two or three times as long, as the fibers become more subdued. The process of paper-making as it has already been described, applies more particularly to papers made from rags. To-day, a very large proportion of the cheaper papers are made from wood, either entirely or in part, and these wood-made papers are subjected to a different treatment, to which further reference will be made.

From the drainer the mass is carted to the beating engine, or "beater," which is very similar in construction to the washer just described. The knives on the roll in the beater are grouped three together instead of two, and are placed nearer the bottom or bed-plate in order to separate more thoroughly the fibers. In the beater are performed many and varying manipulations, designed not only to secure a more perfect product but also to produce different varieties of paper. It is the theory of the beating process that the fibers are not cut, but are drawn out to their utmost extent. In watching the operations of the "beater," one notices on the surface of the slowly revolving mass of fibers, floating bluing, such as the thrifty housewife uses to whiten fine fabrics. This familiar agency of the laundry is introduced into the solution of fibers with the same end in view that is sought in the washtub—to give the clear white color that is so desirable. Many of the inventions and discoveries by which the world has profited largely have been due primarily to some fortunate accident, and according to a pretty story upon which paper-makers have set the seal of their belief for more than one hundred and fifty years, the use of bluing was brought about in the same way. About the year 1746, so runs the story, a Mrs. Buttonshaw, the wife of an English paper-maker, accidentally dropped into a tub of pulp the bag of bluing, or its contents, which she was about to use in a washing of fine linen. Frightened at what she had done and considering it the part of wisdom to keep silence, she discreetly held her peace and awaited results. But when her husband had expressed great wonder and admiration over the paper made from that particular pulp, and had sold it in London at an advance of several shillings over the price of his other paper, which had not met with any such accident, she realized that the time for silence had passed. Her account of the happy accident led her grateful husband to purchase a costly scarlet cloak for her on his next visit to London town. This accident brought about another result which was to prove of inestimable value to the future paper-maker—the use of bluing in paper when especial whiteness is desired.

Important as the bluing or coloring is, however, it is only one of the numerous operations or manipulations that take place in the beater. Many of these, such as engine-sizing and body-coloring, require skill and constant watchfulness. Here, too, if anywhere, adulteration takes place. It is sometimes necessary to secure a fine-appearing paper at small cost, and it is profitable to add to its weight. In such cases a process of "loading" takes place here, and clay or cheap, heavy fibers are added. Clay is of value not only to increase the weight but also to render the paper more opaque, so as to prevent type or illustrations from showing through, while at the same time it makes possible a smoother surface by filling the pores in the paper. But while it adds to the weight, clay must, of necessity, weaken the paper. In engine-sizing, which is done in the beater, the size is thoroughly incorporated with the fibers as these revolve or flow around the engine. This sizing renders the paper more nearly impervious to moisture. The difference between a paper that is sized and that has a repellent surface which prevents the ink from settling into it when it is written upon, and an ordinary blotting-paper with its absorbent surface, is due entirely to the fact that the former is most carefully treated with sizing both in the beating engine and in the size tub or vat referred to later, whereas in the latter paper it is omitted. If the paper is to be tinted or body-colored, colors made from aniline are generally used. Only in the highest grade of writing-paper and in some few papers that demand colors fast to the light is any other order of coloring matter employed. As may be easily imagined, considerable skill is required to secure exactly the desired tint, and to get the coloring matter so evenly mixed that each small fiber shall receive its proper tint, and thus to insure that the paper when finished shall be of uniform color and not present a mottled appearance.

When the operations of the beating engine have been completed, a most interesting process begins which marks a vast advance over the earlier method of forming the sheets of paper with mould and deckel, straining off the water, and shaking the frame with a quick motion to mat the fibers together. The patient striving toward something better which has marked all the centuries since man first learned to carve his rude records, finds its consummation in the process of making paper in a continuous web. This result is accomplished by a machine first invented by Louis Robert, a workman in a mill at Enonnes, France, who obtained a French patent, with a bounty of eight thousand francs for its development. This he later sold to M. Didot, the proprietor of the mill, who crossed the Channel into England, where, with the aid of a skilled mechanic, the machine was in a measure perfected, and then sold to Henry and Sealy Fourdrinier. They, with the further aid of Bryan Donkin, their employee and expert engineer, made many additional improvements, and sank in the enterprise some sixty thousand pounds sterling, for which their only reward was blighted hopes and embittered lives. In 1847 the London *Times* made a fruitless appeal on behalf of the surviving brother, who was eighty years of age and in great poverty. It is seldom that the world voluntarily makes return to those who have bestowed upon it great material or moral benefits, though it is ever ready to expend its treasure for engines of destruction and to magnify and reward those who have been most successful in destroying human life.

The first "machine" mill was started at Frogmore, Hertz, England, in 1803, which was the year of the great Louisiana Purchase by the United States, and it is not difficult to say which event has been productive of the greater and more beneficial results to this nation. Through this invention and its improvements, the modern newspaper and magazine, with their tens and hundreds of thousands of copies daily, have been made possible, and men of all classes have been brought in touch with the best thought of the day. Whatever makes for greater intelligence and enlightenment throughout a nation makes for the greater stability of the national life, and gives new emphasis to Bulwer's words:

Take away the sword; States can be saved without it—bring the pen.

If to-day the power of the pen over the sword is greater than it has ever been before, its increased and increasing influence must be credited in large measure to the inventive genius and the public-spirited enterprise that has made possible the great output of our modern paper-mills. So thoroughly did these forces do their work in the beginning that in the century that has elapsed since the Fourdrinier brothers sacrificed themselves and their means in the perfecting of their machine, there have been really no changes in the fundamental principle. Those that have been made have been in the nature of further development and improvement, such as increasing the speed and widening the web, thereby multiplying the product many fold.

But let us resume the interesting journey of the rags, which had reached a state of purification and perfection as pulp, and which we left in the beaters. In some grades of paper the perfected and prepared pulp is taken from the beaters and passed through what is known as a "refining" or "Jordan" engine for the purpose of more thoroughly separating the fibers and reducing them to extreme fineness. The refining engines are, however, used only in the manufacture of certain grades of paper. The pulp is next taken from the beater or refining engine, as the case may be, to what is called a "stuff-chest," an inclosed vat partly filled with water, in which a contrivance for shaking and shifting, properly called an "agitator," keeps the fibers in suspension.

From the stuff-chest the mixture is pumped into what is known as the "mixing" or "regulating" box. Here the stream first passes over the "sand-tables" in a continuous flow. These are composed of little troughs with cross-pieces, and are covered at the bottom with long-haired felt, to catch any sand or dirt that may still adhere after the numerous operations to which the pulp has been subjected. The flow is then forced through the "screen," which is a horizontal piece of metal pierced with slots. For very fine paper these slots are so small as to be only one one-hundredth of an inch in width. They are usually about a quarter of an inch apart. Through these tiny apertures the fibers must find their way, leaving behind in their difficult passage all lumps, dirt, or knotted fibers which would mar the perfection of the product toward which they are tending. A vibrating motion is given to the screen as the flow passes over it, or revolving strainers may be used.

When the screen has finished its work, the water carrying the pulp in solution flows in an even stream, the volume of which varies according to the width of the web of paper to be produced, through a discharge-cock onto the Fourdrinier or cylinder machine, as the case may be, each of which will be duly described. This stream has a filmy appearance and is of diverse color, depending upon the shade of paper to be produced. From its consistency, which is about that of milk, it is difficult to imagine that it floats separate particles of fiber in such quantities as, when gathered on the wire cloth and passed to a felt blanket and then pressed between rollers, to form in a second of time a broad web of embryo paper sufficiently strong and firm to take definite form. Man's mastery of the process by which this startling and wonderful change is effected has come as one of the rewards of his long and patient study.

The Fourdrinier machine, which preserves at least the name of the enterprising developers of the invention, takes up the work that was formerly done by the molder. The wire cloth upon which the fibers are discharged is an endless belt, the full width of the paper machine. Upon this the fibers spread out evenly, being aided by a fan-shaped rubber or oil cloth, which delivers the smooth stream under a gate regulated to insure perfect evenness and to fix uniformly the fibers of the web now commencing its final formation. Deckel-straps of india-rubber are fastened on both sides of the wire screen, and move with it, thus holding the watery pulp in place. The deckel-straps are adjustable and fix or regulate the width of the paper. These and the gate, or "slicer," are attached to what is termed the deckel-frame, which corresponds to the deckel used by paper-makers in the days when the manufacture was carried on by hand. As the stream flows onto the endless belt of wire cloth, the water which has borne the fibers filters into the trough beneath. Being charged with very fine fibers, size, coloring matter, and other similar ingredients, it is carried back into the pulp-chest to save these materials, as well as to contribute again to the extra supply of water needed. For this reason the trough into which it falls from the revolving "wire" is called the "save-all." A shaking motion is imparted to the "wire" from the frame upon which rest the rolls that keep it in its never-ending round. This aids in draining away the water and mats or interlaces the fibers together. At the end of the "save-all," where the fibers are to leave the "wire" for the next stage of their journey, suction-boxes are placed, provided with an air-pump to take up the surplus water that has not yet found its way through the meshes. Between these suction-boxes above the wire is a wire-covered roll which impresses the newly formed sheet; this impression cylinder is called a "dandy roll," and it is from this that the web receives the markings or impressions that characterize different papers. All watermarks, patterns, and designs which it is desired to have appear in the paper are put upon this roll and here impressed upon the soft sheet, which is clarified and left transparent at the point of contact. Thus the impression is permanently fixed in the fiber, so that it can be seen at any time by holding the sheet to the light. The power of suggestiveness is a quality which is highly esteemed wherever it is found, and which frequently furnishes a standard of judgment.

Judged by such a criterion, the impression cylinder, or "dandy roll," has an added value, for in all probability its operation suggested the idea of printing from cylinders, as in our present web or perfecting presses.

The matted pulp, now having sufficient body, passes on between two rolls covered with felt which deliver the web of damp paper upon an endless belt of moist felt, while the "wire" passes under and back to continue a fresh supply. The paper is as yet too fragile to travel alone, and the web felt carries it between two metal rolls called the first press-rolls. These squeeze out more water, give a greater degree of compactness to the fibers, smooth the upper surface, and finally deliver the web of paper to a second felt apron which carries it under and to the back of the second press-rolls. In this way the under surface comes to the top, and is in its turn subjected to the smoothing process. A delicate scraper or blade, the length of the press-rolls, is so placed on each roll that should the endless web from any cause be broken, the blade may operate with sufficient force to prevent the wet paper from clinging to the rolls and winding about them. From this point the paper travels alone, having become firm and strong enough to sustain its own weight; passing above the second press-rolls, it resumes its onward journey around the drying cylinders, passing over and under and over and under. The drying cylinders are hollow and heated by steam, their temperature being regulated according to requirements. These driers, made from iron or steel, are usually from three to four feet in diameter and vary in length according to the width of the machine. There are from twelve to fifty of these cylinders, their number depending upon the character and weight of the paper to be produced, very heavy sheets requiring many more drying cylinders than sheets of lighter weight.

Strange, almost phenomenal, conditions come about in the transformation from filmy pulp to finished paper. A sheet which, though formed, is at the first press-roll too fragile to carry its own weight, becomes possessed of a final strength and power that is almost incredible. The myriad of minute fibers composing the sheet, upon drying uniformly, possesses great aggregate strength. A sheet of paper yields readily to tearing, but the same sheet, when a perfectly even tension is applied, will demonstrate that it is possessed of wonderful resisting power. In evidence may be cited an instance that seems almost beyond belief. Through some curious mishap a web of heavy paper, in fact, bristol board, which had been thoroughly formed, was suddenly superheated and then cooled while still on the driers. This was caused by a difference in temperature of the driers and resulted in the sudden contraction of the web of bristol; the strain on the machine was so great that not only were the driving-cogs broken on two of the driers around which the paper was at the moment passing, but the driers themselves were actually lifted out of place, showing a resisting power in the paper of at least several tons. The paper now passes to the upright stack of rolls which are known as "calenders." The word is derived from calendra; a corruption of cylindrus, a roller or cylinder. They are simply rollers revolving in contact, and heated from the interior by steam. These calenders are used for giving to the paper a smooth and even surface, and are also employed in the smoothing and finishing of cloth. The speed with which the paper passes through these cylinders is remarkable, from one hundred to five hundred feet running through and over the machine in a minute; and in some of the most recent mills the web is as wide as one hundred and fifty-six inches (thirteen feet); this is very nearly double the average machine width of a very few years ago, while the speed has increased in proportionate ratio; only a few years ago the maximum speed was from two hundred and fifty to three hundred feet per minute; at this writing (1900) there are machines in operation which run as high as five hundred feet per minute. But great as has been the increase in the production of paper, the demand has kept pace steadily. The wonderful product of the rag-bag holds an invincible position in the world's economy.

For machine-finished book and print papers, as well as for other cheaper grades, the process ends with the calenders, after which the paper is slit into required widths by disc-knives which are revolving, and so cut continuously. Paper intended for web newspaper presses is taken off in continuous rolls of the widths required, varying from seventeen to seventy-six inches, according to the size of the paper to be printed. These reels contain from fifteen to twenty-five thousand lineal feet of paper, or from three to five miles. The amount of paper used in disseminating the news of the day is enormous; sometimes one or two mills are required to manufacture the supply for a single metropolitan daily, while one New York newspaper claims to have used four hundred and fifty tons of paper in one Christmas edition, which is about four times the amount of its regular daily consumption.

After having been slit into the proper widths by the revolving knives, ordinary flat and book papers are cut into sheets by a straight knife revolving at proper intervals on a horizontal drum. The paper, in sheets, is carried by a travelling apron to a receiving table at the end of the machine, where the sheets as they fall are carefully examined by experts, usually women, who remove any that may be imperfect.

The entire length of a paper machine, from the screens to the calenders, is about one hundred and twenty-five feet, while the height varies, the average being about ten feet. The machines, while necessarily of the finest adjustment, are ponderous and heavy, weighing in some cases as much as four hundred tons, this being the weight of the machine itself, exclusive of its foundations. The machine-room is of necessity well lighted and thoroughly ventilated, and should be kept clean throughout, as cleanliness is an essential factor in the making of good paper. While the same general process applies to all classes of paper made, the particular character of any paper that is to be produced determines exactly the details of the process through which it shall pass and regulates the deviations to be made from the general operations in order to secure special results. For example, some papers are wanted with a rough or "antique" finish, as it is called; in such cases calendering is omitted. Another special process is that by which the paper is made with a ragged or "deckel-edge;" this result is obtained in some mills by playing a stream of water upon the edge of the pulp, crushing and thinning it, and thus giving it a jagged appearance. At the present time this "deckel-edge" paper is being quite extensively used in high-class bookwork. In the case of writing papers, as has already been stated in the description of the beating engines, a vegetable sizing made from resinous matter is introduced into the paper pulp while it is still in solution, and mixes with it thoroughly, thus filling more or less completely the pores of the pulp fibers. This is found sufficient for all ordinary book-papers, for papers that are to be printed upon in the usual way, and for the cheapest grades of writing-paper, where the requirements are not very exacting and where a curtailment of expense is necessary. For the higher grades of writing-paper, however, a distinctly separate and additional process is required. These papers while on the machine in web form are passed through a vat which is called the size-tub, and which is filled with a liquid sizing made of gelatine from clippings of the horns, hides, and hoofs of cattle, this gelatine or glue being mixed with dissolved alum and made fluid in the vat. Papers which are treated in this way are known as "animal," or "tub-sized."

We have duly described machine-dried papers, but these higher grades of writing-papers are dried by what is known as the loft, or pole-dried process. Such paper is permitted to dry very slowly in a loft specially constructed for the purpose, where it is hung on poles several days, during which time the loft is kept at a temperature of about 100° Fahrenheit.

Another detail of considerable importance is that of the "finish" or surface of the paper. When paper with a particularly high or glossy surface is desired, it is subjected to a separate process, after leaving the paper machine, known as supercalendering.

"Supercalendering" is effected by passing the web through a stack of rolls which are similar to the machine calenders already described. These rolls are composed of metal cylinders, alternating with rolls made of solidified paper or cotton, turned exactly true, the top and bottom rolls being of metal and heavier than the others; a stack of supercalenders is necessarily composed of an odd number of rolls, as seven, nine, or eleven. The paper passes and repasses through these calenders until the requisite degree of smoothness and polish has been acquired. The friction in this machine produces so much electricity that ground wires are often used to carry it off in order that the paper may not become so highly charged as to attract dust or cause the sheets to cling together. When the fine polish has been imparted, the rolls of paper go to the cutting machines, which are automatic in action, cutting regular sheets of the required length as the paper is fed to them in a continuous web. In the manufacture of some high grades of paper, such as linens and bonds, where an especially fine, smooth surface is required, the sheets after being cut are arranged in piles of from twelve to fifteen sheets, plates of zinc are inserted alternately between them, and they are subjected to powerful hydraulic pressure. This process is termed "plating," and is, of course, very much more expensive than the process of supercalendering described above.

From the cutters, the sheets are carried to the inspectors, who are seated in a row along an extended board table before two divisions with partitions ten or twelve inches high, affording spaces for the sheets before and after sorting. The work of inspection is performed by women, who detect almost instantly any blemish or imperfection in the finished product as it passes through their hands. If the paper is to be ruled for writing purposes, it is then taken to the ruling machines, where it is passed under revolving discs or pens, set at regular intervals. These convey the ruling ink to the paper as it passes on through the machine, and thus form true and continuous lines. If the paper is to be folded after ruling, as in the case of fine note-papers, the sheets pass on from the ruling machine to the folding machines, which are entirely automatic in their action. The paper is stacked at the back of the first folding guide and is fed in by the action of small rubber rollers which loosen each sheet from the one beneath, and push it forward until it is caught by the folding apparatus. Man's mechanical ingenuity has given to the machines of his invention something that seems almost like human intelligence, and in the case of the folding machine, the action is so regular and perfect that there seems to be no need of an attendant, save to furnish a constant supply of sheets. The folding completed, cutting machines are again brought into requisition, to cut and trim the sheets to the size of folded note or letter-paper, which is the final operation before they are sent out into the world on their mission of usefulness. The finished paper may or may not have passed through the ruling and folding process, but in either case it goes from the cutters to the wrappers and packers, and then to the shipping-clerks, all of whom perform the duties indicated by their names. The wonderful transformation wrought by the magic wand of science and human invention is complete, and what came into the factory as great bales of offensive rags, disgusting to sight and smell, goes forth as delicate, beautiful, perfected paper, redeemed from filth, and glorified into a high and noble use. Purity and beauty have come from what was foul and unwholesome; the highly useful has been summoned forth from the seemingly useless; a product that is one of the essential factors in the world's progress, and that promises to serve an ever-increasing purpose, has been developed from a material that apparently held not the slightest promise. Well might the *Boston News Letter* of 1769 exclaim in quaint old rhyme:

Rags are as beauties which concealèd lie,But when in paper, charming to the eye!Pray save your rags, new beauties to discover,For of paper truly every one's a lover;By the pen and press such knowledge is displayedAs would not exist if paper was not made.

And well may man pride himself on this achievement, this marvelous transformation, which represents the fruitage of centuries of striving and endeavor!

Up to this point the reference has been almost entirely to paper made from rags, but radical improvements have been made, caused by the introduction of wood pulp, and these are of such importance that the account would not be complete without some mention of them. These changes are mainly in the methods of manipulating the wood to obtain the pulp, for when that is ready, the process from and including the "washers" and "beaters," is very similar to that already described. All papers, whether made from rags or wood, depend upon vegetable fiber for their substance and fundamental base, and it is found that the different fibers used in paper-making, when finally subdued, do not differ, in fact, whether obtained from rags or from the tree growing in the forest. In the latter case the raw wood is subjected to chemical treatment which destroys all resinous and foreign matters, leaving merely the cellular tissue, which, it is found, does not differ in substance from the cell tissue obtained after treating rags. In either case this cellular tissue, through the treatment to which the raw material is subjected, becomes perfectly plastic or moldable, and while the paper made from one differs slightly in certain characteristics from the paper made from the other, they are nevertheless very similar, and it might be safe to predict that further perfecting of processes will eventually make them practically alike.

The woods used for this purpose are principally poplar and spruce, and there are three classes of the wood pulp: (1) mechanical wood, (2) soda process wood, and (3) sulphite wood pulp. The first method was invented in Germany in 1844. The logs are hewn in the forest, roughly barked, and shipped to the factory, where the first operation is to cut them up by steam saws into blocks about two feet in length. Any bark that may still cling to the log is removed by a rapidly revolving corrugated wheel of steel, while the larger blocks are split by a steam splitter. The next stage of their journey takes these blocks to a great millstone set perpendicularly instead of horizontally. Here a very strong and ingenious machine receives one block at a time, and with an automatically elastic pressure holds it sidewise against the millstone, which, like the mills of the gods, "grinds exceeding fine," and with the aid of constantly flowing water rapidly reduces these blocks to a pulpy form. This pulp is carried into tanks, from which it is passed between rollers, which leave it in thick, damp sheets, which are folded up evenly for shipment, or for storage for future use. If a paper-mill is operated in connection with the pulp-mill, the wood pulp is not necessarily rolled out in sheets, but is pumped directly from the tanks to the beaters.

In the preparation of pulp by the other processes, the blocks are first thrown into a chipping machine with great wheels, the short, slanting knives of which quickly cut the blocks into small chips.

In the soda process, invented by M. Meliner in France in 1865, the chips from spruce and poplar logs are boiled under pressure in a strong solution of caustic soda.

When sulphite wood pulp is to be prepared, the chips are conveyed from the chipper into hoppers in the upper part of the building. Here they are thrown into great upright iron boilers or digesters charged with lime-water and fed with the fumes of sulphur which is burned for the purpose in a furnace adjoining the building and which thus forms acid sulphide of lime. The sulphite process was originally invented by a celebrated Philadelphia chemist, but was perfected in Europe.

The "cooking," or boiling, to which the wood is subjected in both the soda and sulphite processes, effects a complete separation of all resinous and foreign substances from the fine and true cell tissue, or cellulose, which is left a pure fiber, ready for use as described. In the case of all fibers, whether rag or wood, painstaking work counts, and the excellence of the paper is largely dependent upon the time and care given to the reduction of the pulp from the original raw material.

Chemical wood pulp of the best quality makes an excellent product, and is largely used for both print and book paper; it is frequently mixed with rag pulp, making a paper that can scarcely be distinguished from that made entirely from fine rags, though it is not of the proper firmness for the best flat or writing papers. All ordinary newspapers, as well as some of the cheaper grades of book and wrapping paper, are made entirely from wood, the sulphite or soda process supplying the fiber, and ground wood being used as a filler. In the average newspaper of to-day's issue, twenty-five per cent of sulphite fiber is sufficient to carry seventy-five per cent of the ground wood filler. The value of the idea is an economical one entirely, as the ground wood employed costs less than any other of the component parts of a print-paper sheet.

The cylinder machine, to which reference was made earlier in the chapter, was patented in 1809 by a prominent paper-maker of England, Mr. John Dickinson. In this machine, a cylinder covered with wire cloth revolves with its lower portion dipping into a vat of pulp, while by suction a partial vacuum is maintained in the cylinder, causing the pulp to cling to the wire until it is conveyed to a covered cylinder, which takes it up and carries it forward in a manner similar to the system already described. This machine is employed in making straw-board and other heavy and cheap grades of paper.

Generous Mother Nature, who supplies man's wants in such bountiful fashion, has furnished on her plains and in her forests an abundance of material that may be transformed into this fine product of human ingenuity. Esparto, a Spanish grass grown in South Africa, has entered largely into the making of print-paper in England. Mixed with rags it makes an excellent product, but the chemicals required to free it from resin and gritty silica are expensive, while the cost of importation has rendered its use in America impractical. Flax, hemp, manila, jute and straw, and of course old paper that has been once used, are extensively employed in this manufacture, the process beginning with the chemical treatment and boiling that are found necessary in the manipulation of rags. The successful use of these materials has met demands that would not otherwise have been supplied. As a result, the price has been so cheapened that the demand for paper has greatly increased, and its use has been extended to many and various purposes.

Many additional items of interest might be described in connection with the methods of manufacturing paper, but as this work is intended for the general reader, rather than for the manufacturer, those wishing further information are referred to technical works on the subject.

The best linen rags are used for the highest grades of writing and bond papers, while ordinary note, letter, and flat papers are made from cotton rags. In some mills, such as the government mill at Dalton, Massachusetts, where the government paper is made for banknotes, and in others where the finest ledger papers are manufactured, none but new, clean rags are used. These come from the remnants left in the making of linen goods. In the government mill where is made the paper for our national currency, or "greenbacks," there is a special attachment on the machine for introducing into the paper the silk threads that are always to be seen in our paper money. This attachment is just above the "wire" on the machine, and consists of a little conducting trough, through which flows, from a receptacle near the machine, a stream of water holding the silk threads in solution. The trough extends across the machine, and is provided at intervals with openings through which the short pieces of silk thread are automatically released, and sprinkled continuously onto the web of pulp as it passes beneath. The paper is thus distinguished, and infringement and possible counterfeiting are made extremely difficult by the fact that the government absolutely forbids the making of paper by others under a similar process, as well as the production of any paper containing these silk threads. The laws of the United States pertaining to anything that borders on infringement of our various money issues, both metal and currency, are most rigid; anything approaching a similarity of impression is prohibited, and a cut, stamp, or impression of any character that approaches in its appearance any money issue of our government is considered a violation of the law against counterfeiting, and is dealt with severely. The government takes the same uncompromising position in regard to the fabrics used in printing its paper-money issues, and it will be quickly seen that the silk thread process described above it is so great a variation from anything required in the mercantile world that it would be difficult to produce a paper at all similar without an ulterior purpose being at once apparent. For this reason the silk thread interspersion is in reality a very effective medium in preventing counterfeiting, not only on account of its peculiar appearance but also because of the elaborate methods necessary in its production.

In those mills making the finest grades of paper, much of the process of thrashing, beating, dusting, and cleaning necessary in the ordinary mill is omitted. The cleanliness and brightness which are reached only at the "washer" and "beater" engines in the process of manufacturing the lower grades of paper from cheaper rags, prevail at every step in these higher grade mills.

One of the first requisites in making good paper, especially the better grades, is an abundance of pure water, and spring-water, where available, is preferred.

The effort has been made in the description given to cover the process of making paper from the crudest rags. In enumerating the several kinds of paper in another chapter, brief reference will be made to the varying methods required in their manufacture. In this chapter, no attempt has been made to cover more than the principal divisions or varieties of paper—writing, print, and wrapping papers.

The United States, with characteristic enterprise, leads the world in paper-making, supplying about one-third of all that is used on the globe. The city of Holyoke, in Massachusetts, is the greatest paper center in the world, turning out each working-day some two hundred tons of paper, nearly one-half of which is "tub-sized," "loft-dried" writings. The region in the vicinity of Holyoke is dotted with paper-mills, and within a few miles of the city is made about one-half of all the "loft-dried" writings produced in the United States. The tiny acorn planted two centuries ago has waxed with the years, gaining strength and vigor with the increasing strength of the nation, till now it has become a giant oak, whose branches extend to the lands beyond the seas.

THE EXPOSITION OF AN IDEA

THE GOSPEL OF RELAXATION

William James

I wish in the following hour to take certain psychological doctrines and show their practical applications to mental hygiene,—to the hygiene of our American life more particularly. Our people, especially in academic circles, are turning towards psychology nowadays with great expectations; and, if psychology is to justify them, it must be by showing fruits in the pedagogic and therapeutic lines.

The reader may possibly have heard of a peculiar theory of the emotions, commonly referred to in psychological literature as the Lange-James theory. According to this theory, our emotions are mainly due to those organic stirrings that are aroused in us in a reflex way by the stimulus of the exciting object or situation. An emotion of fear, for example, or surprise, is not a direct effect of the object's presence on the mind, but an effect of that still earlier effect, the bodily commotion which the object suddenly excites; so that, were this bodily commotion suppressed, we should not so much *feel* fear as call the situation fearful; we should not feel surprise, but coldly recognize that the object was indeed astonishing. One enthusiast has even gone so far as to say that when we feel sorry it is because we weep, when we feel afraid it is because we run away, and not conversely. Some of you may perhaps be acquainted with the paradoxical formula. Now, whatever exaggeration may possibly lurk in this account of our emotions (and I doubt myself whether the exaggeration be very great), it is certain that the main core of it is true, and that the mere giving way to tears, for example, or to the outward expression of an anger-fit, will result for the moment in making the inner grief or anger more acutely felt. There is, accordingly, no better known or more generally useful precept in the moral training of youth, or in one's personal self-discipline, than that which bids us pay primary attention to what we do and express, and not to care too much for what we feel. If we only check a cowardly impulse in time, for example, or if we only *don't* strike the blow or rip out with the complaining or insulting word that we shall regret as long as we live, our feelings themselves will presently be the calmer and better, with no particular guidance from us on their own account. Action seems to follow feeling, but really action and feeling go together; and by regulating the action, which is under the more direct control of the will, we can indirectly regulate the feeling, which is not.

Thus the sovereign voluntary path to cheerfulness, if our spontaneous cheerfulness be lost, is to sit up cheerfully, to look round cheerfully, and to act and speak as if cheerfulness were already there. If such conduct does not make you soon feel cheerful, nothing else on that occasion can. So to feel brave, act as if we *were* brave, use all our will to that end, and a courage-fit will very likely replace the fit of fear. Again, in order to feel kindly toward a person to whom we have been inimical, the only way is more or less deliberately to smile, to make sympathetic inquiries, and to force ourselves to say genial things. One hearty laugh together will bring enemies into a closer communion of heart than hours spent on both sides in inward wrestling with the mental demon of uncharitable feeling. To wrestle with a bad feeling only pins our attention on it, and keeps it still fastened in the mind; whereas, if we act as if from some better feeling, the old bad feeling soon folds its tent like an Arab, and silently steals away.

The best manuals of religious devotion accordingly reiterate the maxim that we must let our feelings go, and pay no regard to them whatever. In an admirable and widely successful little book called *The Christian's Secret of a Happy Life*, by Mrs. Hannah Whitall Smith, I find this lesson on almost every page. *Act* faithfully, and you really have faith, no matter how cold and even how dubious you may feel. "It is your purpose God looks at," writes Mrs. Smith, "not your feelings about that purpose; and your purpose, or will, is therefore the only thing you need attend to.... Let your emotions come or let them go, just as God pleases, and make no account of them either way.... They really have nothing to do with the matter. They are not the indicators of your spiritual state, but are merely the indicators of your temperament or of your present physical condition."

But you all know these facts already, so I need no longer press them on your attention. From our acts and from our attitudes ceaseless inpouring currents of sensation come, which help to determine from moment to moment what our inner states shall be: that is a fundamental law of psychology which I will therefore proceed to assume.

A Viennese neurologist of considerable reputation has recently written about the *Binnenleben,* as he terms it, or buried life of human beings. No doctor, this writer says, can get into really profitable relations with a nervous patient until he gets some sense of what the patient's *Binnenleben* is, of the sort of unuttered inner atmosphere in which his consciousness dwells alone with the secrets of its prison-house. This inner personal tone is what we can't communicate or describe articulately to others; but the wraith and ghost of it, so to speak, are often what our friends and intimates feel as our most characteristic quality. In the unhealthy-minded, apart from all sorts of old regrets, ambitions checked by shames and aspirations obstructed by timidities, it consists mainly of bodily discomforts not distinctly localized by the sufferer, but breeding a general self-mistrust and sense that things are not as they should be with him. Half the thirst for alcohol that exists in the world exists simply because alcohol acts as a temporary anaesthetic and effacer to all these morbid feelings that never ought to be in a human being at all. In the healthy-minded, on the contrary, there are no fears or shames to discover; and the sensations that pour in from the organism only help to swell the general vital sense of security and readiness for anything that may turn up.

Consider, for example, the effects of a well-toned *motor-apparatus,* nervous and muscular, on our general personal self-consciousness, the sense of elasticity and efficiency that results. They tell us that in Norway the life of the women has lately been entirely revolutionized by the new order of muscular feelings with which the use of the *ski,* or long snow-shoes, as a sport for both sexes, has made the women acquainted. Fifteen years ago the Norwegian women were even more than the women of other lands votaries of the old-fashioned ideal of femininity, "the domestic angel," the "gentle and refining influence" sort of thing. Now these sedentary fireside tabby-cats of Norway have been trained, they say, by the snow-shoes into lithe and audacious creatures, for whom no night is too dark or height too giddy, and who are not only saying good-bye to the traditional feminine pallor and delicacy of constitution, but actually taking the lead in every educational and social reform. I cannot but think that the tennis and tramping and skating habits and the bicycle-craze which are so rapidly extending among our dear sisters and daughters in this country are going also; to lead to a sounder and heartier moral tone, which will send its tonic breath through all our American life.

I hope that here in America more and more the ideal of the well-trained and vigorous body will be maintained neck by neck with that of the well-trained and vigorous mind as the two coequal halves of the higher education for men and women alike. The strength of the British Empire lies in the strength of character of the individual Englishman, taken all alone by himself. And that strength, I am persuaded, is perennially nourished and kept up by nothing so much as by the national worship, in which all classes meet, of athletic outdoor life and sport.

I recollect, years ago, reading a certain work by an American doctor on hygiene and the laws of life and the type of future humanity. I have forgotten its author's name and its title, but I remember well an awful prophecy that it contained about the future of our muscular system. Human perfection, the writer said, means ability to cope with the environment; but the environment will more and more require mental power from us, and less and less will ask for bare brute strength. Wars will cease, machines will do all our heavy work, man will become more and more a mere director of nature's energies, and less and less an exerter of energy on his own account. So that, if the *homo sapiens* of the future can only digest his food and think, what need will he have of well-developed muscles at all? And why, pursued this writer, should we not even now be satisfied with a more delicate and intellectual type of beauty than that which pleased our ancestors? Nay, I have heard a fanciful friend make a still further advance in this "new-man" direction.

With our future food, he says, itself prepared in liquid form from the chemical elements of the atmosphere, pepsinated or half-digested in advance, and sucked up through a glass tube from a tin can, what need shall we have of teeth, or stomachs even? They may go, along with our muscles and our physical courage, while, challenging even more and more our proper admiration, will grow the gigantic domes of our crania, arching over our spectacled eyes, and animating our flexible little lips to those floods of learned and ingenious talk which will constitute our most congenial occupation.

I am sure that your flesh creeps at this apocalyptic vision. Mine certainly did so; and I cannot believe that our muscular vigor will ever be a superfluity. Even if the day ever dawns in which it will not be needed for fighting the old heavy battles against Nature, it will still always be needed to furnish the background of sanity, serenity, and cheerfulness to life, to give moral elasticity to our disposition, to round off the wiry edge of our fretfulness, and make us good-humored and easy to approach. Weakness is too apt to be what the doctors call irritable weakness. And that blessed internal peace and confidence, that *acquiescentia in seipso*, as Spinoza used to call it, that wells up from every part of the body of a muscularly well-trained human being, and soaks the indwelling soul of him with satisfaction, is, quite apart from every consideration of its mechanical utility, an element of spiritual hygiene of supreme significance.

And now let me go a step deeper into mental hygiene, and try to enlist your insight and sympathy in a cause which I believe is one of paramount patriotic importance to us Yankees. Many years ago a Scottish medical man, Dr. Clouston, a mad-doctor as they call him there, or what we should call an asylum physician (the most eminent one in Scotland), visited this country, and said something that has remained in my memory ever since. "You Americans," he said, "wear too much expression on your faces. You are living like an army with all its reserves engaged in action. The duller countenances of the British population betoken a better scheme of life. They suggest stores of reserved nervous force to fall back upon, if any occasion should arise that requires it. This inexcitability, this presence at all times of power not used, I regard," continued Dr. Clouston, "as the great safeguard of our British people. The other thing in you gives me a sense of insecurity, and you ought somehow to tone yourselves down. You really do carry too much expression, you take too intensely the trivial moments of life."

Now Dr. Clouston is a trained reader of the secrets of the soul as expressed upon the countenance, and the observation of his which I quote seems to me to mean a great deal. And all Americans who stay in Europe long enough to get accustomed to the spirit, that reigns and expresses itself there, so unexcitable as compared with ours, make a similar observation when they return to their native shores. They find a wild-eyed look upon their compatriots' faces, either of too desperate eagerness and anxiety or of too intense responsiveness and good-will. It is hard to say whether the men or the women show it most. It is true that we do not all feel about it as Dr. Clouston felt. Many of us, far from deploring it, admire it. We say: "What intelligence it shows! How different from the stolid cheeks, the codfish eyes, the slow, inanimate demeanor we have been seeing in the British Isles!" Intensity, rapidity, vivacity of appearance, are indeed with us something of a nationally accepted ideal; and the medical notion of "irritable weakness" is not the first thing suggested by them to our mind, as it was to Dr. Clouston's. In a weekly paper not very long ago I remember reading a story in which, after describing the beauty and interest of the heroine's personality, the author summed up her charms by saying that to all who looked upon her an impression as of "bottled lightning" was irresistibly conveyed.

Bottled lightning, in truth, is one of our American ideals, even of a, young girl's character! Now it is most ungracious, and it may seem to some persons unpatriotic, to criticise in public the physical peculiarities of one's own people, of one's own family, so to speak. Besides, it may be said, and said with justice, that there are plenty of bottled-lightning temperaments in other countries, and plenty of phlegmatic temperaments here; and that, when all is said and done, the more or less of tension about which I am making such a fuss is a small item in the sum total of a nation's life, and not worth solemn treatment at a time when agreeable rather than disagreeable things should be talked about. Well, in one sense the more or less of tension in our faces and in our unused muscles *is* a small thing: not much mechanical work is done by these contractions. But it is not always the material size of a thing that measures its importance: often it is its place and function. One of the most philosophical remarks I ever heard made was by an unlettered workman who was doing some repairs at my house many years ago. "There is very little difference between one man and another," he said, "when you go to the bottom of it. But what little there is, is very important." And the remark certainly applies to this case. The general over-contraction may be small when estimated in foot-pounds, but its importance is immense on account of its *effects on the over-contracted person's spiritual life.* This follows as a necessary consequence from the theory of our emotions to which I made reference at the beginning of this article. For by the sensations that so incessantly pour in from the over-tense excited body the over-tense and excited habit of mind is kept up; and the sultry, threatening, exhausting, thunderous inner atmosphere never quite clears away. If you never wholly give yourself up to the chair you sit in, but always keep your leg- and body-muscles half contracted for a rise; if you breathe eighteen or nineteen instead of sixteen times a minute, and never quite breathe out at that,—what mental mood *can* you be in but one of inner panting and expectancy, and how can the future and its worries possibly forsake your mind? On the other hand, how can they gain admission to your mind if your brow be unruffled, your respiration calm and complete, and your muscles all relaxed?

Now what is the cause of this absence of repose, this bottled-lightning quality in us Americans? The explanation of it that is usually given is that it comes from the extreme dryness of our climate and the acrobatic performances of our thermometer, coupled with the extraordinary progressiveness of our life, the hard work, the railroad speed, the rapid success, and all the other things we know so well by heart. Well, our climate is certainly exciting, but hardly more so than that of many parts of Europe, where nevertheless no bottled-lightning girls are found. And the work done and the pace of life are as extreme in every great capital of Europe as they are here. To me both of these pretended causes are utterly insufficient to explain the facts.

To explain them, we must go not to physical geography, but to psychology and sociology. The latest chapter both in sociology and in psychology to be developed in a manner that approaches adequacy is the chapter on the imitative impulse. First Bagehot, then Tarde, then Royce and Baldwin here, have shown that invention and imitation, taken together, form, one may say, the entire warp and woof of human life, in so far as it is social. The American over-tension and jerkiness and breathlessness and intensity and agony of expression are primarily social, and only secondarily physiological, phenomena. They are *bad habits*, nothing more or less, bred of custom and example, born of the imitation of bad models and the cultivation of false personal ideals. How are idioms acquired, how do local peculiarities of phrase and accent come about? Through an accidental example set by some one, which struck the ears of others, and was quoted and copied till at last every one in the locality chimed in. Just so it is with national tricks of vocalization or intonation, with national manners, fashions of movement and gesture, and habitual expressions of face. We, here in America, through following a succession of pattern-setters whom it is now impossible to trace, and through influencing each other in a bad direction, have at last settled down collectively into what, for better or worse, is our own characteristic national type,—a type with the production of which, so far as these habits go, the climate and conditions have had practically nothing at all to do.

This type; which we have thus reached by our imitativeness, we now have fixed upon us, for better or worse. Now no type can be *wholly* disadvantageous; but, so far as our type follows the bottled-lightning fashion, it cannot be wholly good. Dr. Clouston was certainly right in thinking that eagerness, breathlessness, and anxiety are not signs of strength: they are signs of weakness and of bad co-ordination. The even forehead, the slab-like cheek, the codfish eye, may be less interesting for the moment; but they are more promising signs than intense expression is of what we may expect of their possessor in the long run. Your dull, unhurried worker gets over a great deal of ground, because he never goes backward or breaks down. Your intense, convulsive worker breaks down and has bad moods so often that you never know where he may be when you most need his help,—he may be having one of his "bad days." We say that so many of our fellow-countrymen collapse, and have to be sent abroad to rest their nerves, because they work so hard. I suspect that this is an immense mistake. I suspect that neither the nature nor the amount of our work is accountable for the frequency and severity of our breakdowns, but that their cause lies rather in those absurd feelings of hurry and having no time, in that breathlessness and tension, that anxiety of feature and that solicitude for results, that lack of inner harmony and ease, in short, by which with us the work is so apt to be accompanied, and from which a European who should do the same work would nine times out of ten be free. These perfectly wanton and unnecessary tricks of inner attitude and outer manner in us, caught from the social atmosphere, kept up by tradition, and idealized by many as the admirable way of life, are the last straws that break the American camel's back, the final overflowers of our measure of wear and tear and fatigue.

The voice, for example, in a surprisingly large number of us has a tired and plaintive sound. Some of us are really tired (for I do not mean absolutely to deny that our climate has a tiring quality); but far more of us are not tired at all, or would not be tired at all unless we had got into a wretched trick of feeling tired, by following the prevalent habits of vocalization and expression. And if talking high and tired, and living excitedly and hurriedly, would only enable us to *do* more by the way, even while breaking us down in the end, it would be different. There would be some compensation, some excuse, for going on so. But the exact reverse is the case. It is your relaxed and easy worker, who is in no hurry, and quite thoughtless most of the while of consequences, who is your efficient worker; and tension and anxiety, and present and future, all mixed up together in our mind at once, are the surest drags upon steady progress and hindrances to our success. My colleague, Professor Münsterberg, an excellent observer, who came here recently, has written some notes on America to German papers. He says in substance that the appearance of unusual energy in America is superficial and illusory, being really due to nothing but the habits of jerkiness and bad co-ordination for which we have to thank the defective training of our people. I think myself that it is high time for old legends and traditional opinions to be changed; and that, if any one should begin to write about Yankee inefficiency and feebleness, and inability to do anything with time except to waste it, he would have a very pretty paradoxical thesis to sustain, with a great many facts to quote, and a great deal of experience to appeal to in its proof.

Well, my friends, if our dear American character is weakened by all this over-tension,—and I think, whatever reserves you may make, that you will agree as to the main facts,—where does the remedy lie? It lies, of course, where lay the origins of the disease. If a vicious fashion and taste are to blame for the thing, the fashion and taste must be changed. And, though it is no small thing to inoculate seventy millions of people with new standards, yet, if there is to be any relief, that will have to be done. We must change ourselves from a race that admires jerk and snap for their own sakes, and looks down upon low voices and quiet ways as dull, to one that, on the contrary, has calm for its ideal, and for their own sakes loves harmony, dignity, and ease.

So we go back to the psychology of imitation again. There is only one way to improve ourselves, and that is by some of us setting an example which the others may pick up and imitate till the new fashion spreads from east to west. Some of us are in more favorable positions than others to set new fashions. Some are much more striking personally and imitable, so to speak. But no living person is sunk so low as not to be imitated by somebody. Thackeray somewhere says of the Irish nation that there never was an Irishman so poor that he didn't have a still poorer Irishman living at his expense; and, surely, there is no human being whose example doesn't work contagiously in *some* particular. The very idiots at our public institutions imitate each other's peculiarities. And, if you should individually achieve calmness and harmony in your own person, you may depend upon it that a wave of imitation will spread from you, as surely as the circles spread outward when a stone is dropped into a lake.

Fortunately, we shall not have to be absolute pioneers. Even now in New York they have formed a society for the improvement of our national vocalization, and one perceives its machinations already in the shape of various newspaper paragraphs intended to stir up dissatisfaction with the awful thing that it is. And, better still than that, because more radical and general, is the gospel of relaxation, as one may call it, preached by Miss Annie Payson Call, of Boston, in her admirable little volume called *Power Through Repose*, a book that ought to be in the hands of every teacher and student in America of either sex. You need only be followers, then, on a path already opened up by others. But of one thing be confident: others still will follow you.

And this brings me to one more application of psychology to practical life, to which I will call attention briefly, and then close. If one's example of easy and calm ways is to be effectively contagious, one feels by instinct that the less voluntarily one aims at getting imitated, the more unconscious one keeps in the matter, the more likely one is to succeed. *Become the imitable thing,* and you may then discharge your minds of all responsibility for the imitation. The laws of social nature will take care of that result. Now the psychological principle on which this precept reposes is a law of very deep and widespread importance in the conduct of our lives, and at the same time a law which we Americans most grievously neglect. Stated technically, the law is this: that *strong feeling about one's self tends to arrest the free association of one's objective ideas and motor processes.* We get the extreme example of this in the mental disease called melancholia.

A melancholic patient is filled through and through with intensely painful emotion about himself. He is threatened, he is guilty, he is doomed, he is annihilated, he is lost. His mind is fixed as if in a cramp on these feelings of his own situation, and in all the books on insanity you may read that the usual varied flow of his thoughts has ceased. His associative processes, to use the technical phrase, are inhibited; and his ideas stand stock-still, shut up to their one monotonous function of reiterating inwardly the fact of the man's desperate estate. And this inhibitive influence is not due to the mere fact that his emotion is *painful.* Joyous emotions about the self also stop the association of our ideas. A saint in ecstasy is as motionless and irresponsive and one-idea'd as a melancholiac. And, without going as far as ecstatic saints, we know how in every one a great or sudden pleasure may paralyze the flow of thought. Ask young people returning from a party or a spectacle, and all excited about it, what it was. "Oh, it was *fine!* it was *fine!* it was *fine!*" is all the information you are likely to receive until the excitement has calmed down. Probably every one of my hearers has been made temporarily half-idiotic by some great success or piece of good fortune. "*Good!* GOOD! GOOD!" is all we can at such times say to ourselves until we smile at our own very foolishness.

Now from all this we can draw an extremely practical conclusion. If, namely, we wish our trains of ideation and volition to be copious and varied and effective, we must form the habit of freeing them from the inhibitive influence of reflection upon them, of egoistic pre-occupation about their results. Such a habit, like other habits, can be formed. Prudence and duty and self-regard, emotions of ambition and emotions of anxiety, have, of course, a needful part to play in our lives. But confine them as far as possible to the occasions when you are making your general resolutions and deciding on your plan of campaign, and keep them out of the details. When once a decision is reached and execution is the order of the day, dismiss absolutely all responsibility and care about the outcome. *Unclamp,* in a word, your intellectual and practical machinery, and let it run free; and the service it will do you will be twice as good. Who are the scholars who get "rattled" in the recitation-room? Those who think of the possibilities of failure and feel the great importance of the act. Who are those who do recite well? Often those who are most indifferent. *Their* ideas reel themselves out of their memory of their own accord. Why do we hear the complaint so often that social life in New England is either less rich and expressive or more fatiguing than it is in some other parts of the world? To what is the fact, if fact it be, due unless to the over-active conscience of the people, afraid of either saying something too trivial and obvious, or something insincere, or something unworthy of one's interlocutor, or something in some way or other not adequate to the occasion? How can conversation possibly steer itself through such a sea of responsibilities and inhibitions as this? On the other hand, conversation does flourish and society is refreshing, and neither dull on the one hand nor exhausting from its efforts on the other, wherever people forget their scruples and take the brakes off their hearts, and let their tongues wag as automatically and irresponsibly as they will.

They talk much in pedagogic circles to-day about the duty of the teacher to prepare for every lesson in advance. To some extent this is useful. But we Yankees are assuredly not those to whom such a general doctrine should be preached. We are only too careful as it is. The advice I should give to most teachers would be in the words of one who is herself an admirable teacher. Prepare yourself in the *subject so well that it shall be always on tap*: then in the class-room trust your spontaneity and fling away all further care.

My advice to students, especially to girl-students, would be somewhat similar. Just as a bicycle-chain may be too tight, so may one's carefulness and conscientiousness be so tense as to hinder the running of one's mind. Take, for example, periods when there are many successive days of examination pending. One ounce of good nervous tone in an examination is worth many pounds of anxious study for it in advance. If you want really to do your best at an examination, fling away the book the day before, say to yourself, "I won't waste another minute on this miserable thing, and I don't care an iota whether I succeed or not." Say this sincerely and feel it; and go out and play, or go to bed and sleep, and I am sure the results next day will encourage you to use the method permanently. I have heard this advice given to a student by Miss Call, whose book on muscular relaxation I quoted a moment ago. In her later book, entitled *As a Matter of Course*, the gospel of moral relaxation, of dropping things from the mind, and not "caring," is preached with equal success. Not only our preachers, but our friends the theosophists and mind-curers of various religious sects are also harping on this string. And with the doctors, the Delsarteans, the various mind-curing sects, and such writers as Mr. Dresser, Prentice Mulford, Mr. Horace Fletcher, and Mr. Trine to help, and the whole band of schoolteachers and magazine-readers chiming in, it really looks as if a good start might be made in the direction of changing our American mental habit into something more indifferent and strong.

Worry means always and invariably inhibition of associations and loss of effective power. Of course, the sovereign cure for worry is religious faith; and this, of course, you also know. The turbulent billows of the fretful surface leave the deep parts of the ocean undisturbed, and to him who has a hold on vaster and more permanent realities the hourly vicissitudes of his personal destiny seem relatively insignificant things. The really religious person is accordingly unshakable and full of equanimity, and calmly ready for any duty that the day may bring forth. This is charmingly illustrated by a little work with which I recently became acquainted, "The Practice of the Presence of God, the Best Ruler of a Holy Life, by Brother Lawrence, being Conversations and Letters of Nicholas Herman of Lorraine, Translated from the French."[7] I extract a few passages, the conversations being given in indirect discourse. Brother Lawrence was a Carmelite friar, converted at Paris in 1666. "He said that he had been footman to M. Fieubert, the Treasurer, and that he was a great awkward fellow, who broke everything. That he had desired to be received into a monastery, thinking that he would there be made to smart for his awkwardness and the faults he should commit, and so he should sacrifice to God his life, with its pleasures; but that God had disappointed him, he having met with nothing but satisfaction in that state....

"That he had long been troubled in mind from a certain belief that he should be damned; that all the men in the world could not have persuaded him to the contrary; but that he had thus reasoned with himself about it: *I engaged in a religious life only for the love of God, and I have endeavored to act only for Him; whatever becomes of me, whether I be lost or saved, I will always continue to act purely for the love of God. I shall have this good at least, that till death I shall have done all that is in me to love Him* ... That since then he had passed his life in perfect liberty and continual joy.

"That when an occasion of practicing some virtue offered, he addressed himself to God, saying, 'Lord, I cannot do this unless Thou enablest me'; and that then he received strength more than sufficient. That, when he had failed in his duty, he only confessed his fault, saying to God, 'I shall never do otherwise, if You leave me to myself: it is You who must hinder my failing, and mend what is amiss.' That after this he gave himself no further uneasiness about it.

"That he had been lately sent into Burgundy to buy the provision of wine for the society, which was a very unwelcome task for him, because he had no turn for business, and because he was lame, and could not go about the boat but by rolling himself over the casks. That, however, he gave himself no uneasiness about it, nor about the purchase of the wine. That he said to God, 'It was his business he was about,' and that he afterward found it well performed. That he had been sent into Auvergne, the year before, upon the same account; that he could not tell how the matter passed, but that it proved very well.

"So, likewise, in his business in the kitchen (to which he had naturally a great aversion), having accustomed himself to do everything there for the love of God, and with prayer upon all occasions, for his grace to do his work well, he had found everything easy during fifteen years that he had been employed there.

"That he was very well pleased with the post he was now in, but that he was as ready to quit that as the former, since he was always pleasing himself in every condition, by doing little things for the love of God.

"That the goodness of God assured him He would not forsake him utterly, and that He would give him strength to bear whatever evil He permitted to happen to him; and, therefore, that he feared nothing, and had no occasion to consult with anybody about his state. That, when he had attempted to do it, he had always come away more perplexed."

The simple-heartedness of the good Brother Lawrence, and the relaxation of all unnecessary solicitudes and anxieties in him is a refreshing spectacle.

The need of feeling responsible all the livelong day has been preached long enough in our New England. Long enough exclusively, at any rate,—and long enough to the female sex. What our girl-students and women-teachers most need nowadays is not the exacerbation, but rather the toning-down of their moral tensions. Even now I fear that some one of my fair hearers may be making an undying resolve to become strenuously relaxed, cost what it will, for the remainder of her life. It is needless to say that that is not the way to do it. The way to do it, paradoxical as it may seem, is genuinely not to care whether you are doing it or not. Then, possibly, by the grace of God, you may all at once find that you *are* doing it, and, having learned what the trick feels like, you may (again by the grace of God) be enabled to go on.

And that something like this may be the happy experience of all my hearers is, in closing, my most earnest wish.

SCIENCE AND RELIGION

Charles Proteus Steinmetz

The problem of religion—that is, of the relations of man with the supernatural, with God and immortality, with the soul, our personality or the ego, and its existence or nonexistence after death—is the greatest and deepest which ever confronted mankind. In the present state of human knowledge, science can give no definite and final conclusions on these subjects, because of the limitations inherent in science.

We must realize that all our knowledge and information and the entire structure of science are ultimately derived from the perceptions of our senses and thereby limited in the same manner and to the same extent as our sense perceptions and our intellect are limited. The success or failure of scientific achievement largely depends on the extent to which we can abstract—that is, make our observations and conclusions independent of the limitations of the human mind. But there are limitations inherent in the human mind beyond which our intellect cannot reach, and therefore science does not and cannot show us the world as it actually is, with its true facts and laws, but only as it appears to us within the inherent limitations of the human mind.

The greatest limitation of the human mind is that all its perceptions are finite, and our intellect cannot grasp the conception of infinity. The same limitation therefore applies to the world as it appears to our reasoning intellect, and in the world of science there is no infinity, and conceptions such as God and the immortality of the ego are beyond the realm of empirical science. Science deals only with finite events in finite time and space, and the farther we pass onward in space or time, the more uncertain becomes the scientific reasoning, until, in trying to approach the infinite, we are lost in the fog of unreasonable contradiction, "beyond science"—that is, "transcendental".

Thus, we may never know and understand the infinite, whether in nature, in the ultimate deductions from the laws of nature in time and in space, or beyond nature, on such transcendental conceptions as God and immortality. But we may approach these subjects as far as the limitations of our mind permit, reach the border line beyond which we cannot go, and so derive some understanding of how far these subjects may appear nonexisting or unreasonable, merely because they are beyond the limitations of our intellect.

There appear to me two promising directions of approach—first, from the complex of thought and research, which in physics has culminated in the theory of relativity; and, second, in a study of the gaps found in the structure of empirical science and what they may teach us.

All events of nature occur in space and in time. Whatever we perceive, whatever record we receive through our senses, always is attached to, and contained in, space and time. But are space and time real existing things? Have they an absolute reality outside of our mind, as a part or framework of nature, as entities—that is, things that are? Or are they merely a conception of the human mind, a form given by the character of our mind to the events of nature—that is, to the hypothetical cause of our sense perceptions? Kant, the greatest and most critical of all philosophers, in his *Critique of Pure Reason (Kritik der Reinen Vernunft)*, concludes that space and time have no absolute existence, but are categories—that is, forms in which the human mind conceives his relation to nature. The same idea is expressed by the poet-philosopher Goethe in his dramatic autobiography *Faust* (in the second part), when he refers to the "Mütter," to the marriage of Achilles and Helena "outside of all time." It is found in ancient time. So Revelation speaks of "there should be time no longer" (ὅτι χρόνος οὐκέτι ἔσται

The work of the great mathematicians of the nineteenth century—Gauss, Riemann, Lobatschefsky, Bolyai—offered further evidence that space is not an empirical deduction from nature, but a conception of the mind, by showing that various forms of space can be conceived, differing from one another and from the form in which the mind has cast the events of nature (the "Euclidean" space). Finally, physical science, in the theory of relativity, has deduced the same conclusions: space and time do not exist in nature by themselves, as empty space and empty time, but their existence is only due to things and events as they occur in nature. They are relative in the relation between us and the events of nature, so much so that they are not fixed and invariable in their properties, but depend upon the observer and the conditions of observation.

We can get an idea of how utterly our perception of nature depends on the particular form of our time conception by picturing to ourselves how nature would look if our time perception were 100,000 times faster, or 100,000 times slower.

In the first case, with our sense perceptions 100,000 times faster, all events in nature would appear to us 100,000 times slower. This would then be a stationary and immovable world. The only motion which we could see with our eyes would be that of the cannon ball, which would crawl slowly along, at less than a snail's pace. The express train going at sixty miles per hour would appear to stand still, and deliberate experiment be required to discover its motion. By noting its position on the track, and noting it again after a period of time as long as five minutes appears to us now, we should find its position changed by three inches. It would be a dangerous world, as there would be many objects—not distinguishable to the senses from other harmless objects—contact with which would be dangerous, even fatal; and one and the same object (as the express train) might sometimes be harmless (when at rest), sometimes dangerous (when in motion), without our senses being able to see any difference.

On the other hand, with our sense perceptions 100,000 times slower, all events in nature would appear to us to occur 100,000 times faster. There would be little rest in nature, and we should see plants, and even stones, move. We should observe, in a period of time not longer than a minute or two appear to us now, a plant start from seed, grow up, flower, bring fruit, and die. Sun and moon would be luminous bands traversing the sky; day and night alternate seconds of light and darkness. Much of nature, all moving things, would be invisible to us. If I moved my arm, it would disappear, to reappear again when I held it still. It would be a usual occurrence to have somebody suddenly appear and just as suddenly disappear from our midst, or to see only a part of a body. The vanishing and the appearance of objects would be common occurrences in nature; and we should speak of "vanishing" and "appearing," instead of "moving" and "stopping." Collisions, usually harmless, with invisible objects would be common occurrences.

As seen, nature and its laws would appear to us very different from what we find them now, with our present time perception.

Thus philosophy, mathematics, and physical science agree that space and time cannot be entities, but are conceptions of the human mind in its relation to nature. But what does this mean, and what conclusions follow from it?

The space of our conception is three-dimensional—that is, extended in three directions. For instance, the north-south direction, the east-west direction, and the up-down direction. Any place or "point" in space thus is located, relative to some other point, by giving its three distances from the latter, in three (arbitrarily chosen) directions.

Time has only one dimension—that is, extends in one direction only, from the past to the future—and a moment or "point" in time thus is located, with reference to another point in time, by one time distance.

But there is a fundamental difference between our space conception and our time conception, in that we can pass through time only in one direction, from the past to the future, while we can pass through space in any direction, from north to south, as well as from south to north—that is, time is irreversible, flows uniformly in one direction, while space is reversible, can be traversed in any direction. This means that when we enter a thing in space, as a house, we can approach it, pass through it, leave it, come back to it, and the thing therefore appears permanent to us, and we know, even when we have left the house and do not see it any more, that it still exists, and that we can go back to it again and enter it. Not so with time. On approaching a thing in time, an event such as a human life, it extends from a point in time—birth—over a length of time—the life—to an end point in time—death—just as the house in space extends from a point in space—say the north wall—over a length of space—its extent—to an end point in space—say the south wall. But when we pass beyond the end point of an event in time—the death of a life—we cannot go back to the event any more; the event has ceased, ended, the life is extinct.

But let us imagine that the same irreversibility applied to the conception of space—that is, that we could move through space only from north to south, and not in the opposite direction. Then a thing in space, as a house, would not exist for us until we approached it. When we were approaching it, it would first appear indistinctly, and more and more distinctly the nearer we approached it, just as an event in time does not exist until we reach the point of its beginning, but may appear in anticipation, in time perspective, when we approach it, the more distinctly, the closer we approach it, until we reach the threshold of the time span covered by the event, and the event begins to exist, the life is born. So to us, if we could move only from north to south, the house would begin to exist only when we reached its north door. That point would be the "birth" of the house. Passing through the span of space covered by the house—this would for us be its existence, its "life," and when we stepped out of the south door the house would cease to exist for us, we could never enter it and turn back to it again—that is, it would be dead and extinct, just as the life when we pass beyond its end point in time. Thus birth and death, appearance and extinction of an event in time, as our life, are the same as the beginning and end point of a thing in space, like a house. But the house appears to us to exist permanently, whether we are in it, within the length between beginning and end point, or not; while the event in time, our life, appears to us to exist only during the length of time when we are between its beginning and its end point in time, and before and after it does not exist for us, because we cannot go back to it or ahead into it. But assume time were reversible, like space—that is, we could go through it in any direction. There would then be no such thing as birth or origin, and death or extinction, but our life would exist permanently, as a part or span of time, just as the house exists as a part or section of space, and the question of immortality, of extinction or nonextinction by death, would then be meaningless. We should not exist outside of the span of time covered by our life, just as we do not exist outside of the part of space covered by our body in space, and to reach an event, as our life, we should have to go to the part of space and to the part of time where it occurs; but there would be no more extinction of the life by going beyond its length in time as there is extinction of a house by going outside of its door, and everything, like a human being, would have four extensions or dimensions—three extensions in space and one in time.

If space and time, and therefore the characteristics of space and time, are not real things or entities, but conceptions of the human mind, then those transcendental questions, as that of immortality after death and existence before birth, are not problems of fact in nature or outside of nature, but are meaningless, just as the question whether a house exists for an observer outside of the space covered by it. In other words, the questions of birth and death, of extinction or immortality, are merely the incidental results of the peculiarity of our conceptions of time, the peculiarity that the time of our conceptions is irreversible, flows continuously at a uniform rate in the same direction from the past to the future.

But if time has no reality, is not an existing entity, then these transcendental problems resulting from our time conception, of extinction or immortality, have no real existence, but are really phenomena of the human mind, and cease to exist if we go beyond the limitations of our mind, beyond our peculiar time conception.

It is interesting to realize that the modern development of science, in the relativity theory, has proved not only that time is not real, but a conception, but also has proved that the time of our conception does not flow uniformly at constant rate from past to future, but that the rate of the flow of time varies with the conditions; the rate of time flow of an event slows down with the motion relative to the event.

But the conception of a reversal of the flow of time is no more illogical than the conception of a change of the rate of the flow of time. It is inconceivable, because it is beyond the limitations of our mind.

Thus we see that the questions of life and death, of extinction and immortality, are not absolute problems, but merely the result of the limitations of our mind in its conception of time, and have no existence outside of us.

After all, to some extent we conceive time as reversible, in the conception of historical time. In history we go back in time at our will, and traverse with the mind's eye the times of the past, and we then find that death and extinction do not exist in history, but the events of history, the lives of those who made history, exist just as much outside of the span of time of their physiological life—that is, are immortal in historical time. They may fade and become more indistinct with the distance in time, just as things in space become more indistinct with the distance in space, but they can be brought back to full clearness and distinction by again approaching the things and events, the former moving through space, the latter moving through the historical time—that is, by looking up and studying the history of the time.

THE ENTITY "X"

Scientifically, life is a physico-chemical process. Transformations of matter, with which the chemist deals, and transformations of energy, with which the physicist deals, are all that is comprised in the phenomenon of life; and mind, intellect, soul, personality, the ego, are mere functions of the physico-chemical process of life, vanishing when this process ceases, but are not a part of the transformations of matter and of energy. If you thus speak of "mental energy," it scientifically is a misnomer, and mind is not energy in the physical sense. It is true that mental effort, intellectual work, is accompanied by transformations of matter, chemical changes in the brain, and by transformations of energy. But the mental activity is not a part of the energy or of the matter which is transformed, but the balance of energy and of matter closes.

In the energy transformations accompanying mental activity, just as much energy of one form appears as energy of some other form is consumed, and the mental activity is no part of the energy. In the transformations of matter accompanying mental activity, just as much matter of one form appears as matter of some other form is consumed, and the mental activity is no part of either—that is, neither energy nor matter has been transformed into mental activity, nor has energy or matter been produced by mental activity. All attempts to account for the mental activity as produced by the expenditure of physical energy, or as producing physical energy—that is, exerting forces and action—have failed and must fail, and so must any attempt to record or observe and measure mental activity by physical methods—that is, methods sensitive to the action of physical forces.

But what, then, is mind? Is it a mere phenomenon, accompanying the physico-chemical reactions of life and vanishing with the end of the reaction, just as the phenomenon of a flame may accompany a chemical reaction, and vanish when the reaction is completed? Or is mind an entity, just like the entity energy and the entity matter, but differing from either of them—in short, a third entity? We have compared mind with the phenomenon of a flame accompanying a chemical reaction; but, after all, the flame is not a mere phenomenon, but is an entity, is energy.

More than once, in the apparently continuous and unbroken structure of science, wide gaps have been discovered into which new sections of knowledge fitted, sections the existence of which had never been suspected. So in Mendelejeff's *Periodic System of the Elements* all chemical elements fitted in without gaps—in a continuous series (except a few missing links, which were gradually discovered and filled in). Nevertheless, the whole group of six noble gases, from helium to emanium, were discovered and fitted into the periodic system at a place where nobody had suspected a gap.

One of the most interesting of such unsuspected gaps in the structure of science is the following, because of its pertinency to the subject of our discussion.

In studying the transformations of matter, the chemist records them by equations of the form:

(1) $2H_2 + O_2 = 2H_2O$, which means:

Two gram molecules of hydrogen H_2(2 X 2 = 4 grams) and 1 gram molecule of oxygen O_2(1 X 32 grams), combine to 2 gram molecules of water vapor H_2O (2 X 18 = 36 grams).

For nearly a hundred years chemists wrote and accepted this equation; innumerable times it has been experimentally proved by combining 4 parts of hydrogen and 32 parts of oxygen to 36 parts of water vapor; so that this chemical equation would appear as correct and unquestionable as anything can be.

Nevertheless, it is wrong, or rather incomplete. It does not give the whole event, but omits an essential part of it, and now we write it:

(2) $2H_2 + O_2 = 2H_2O$ + 293,000 J., which means:

The matter *and energy* of 2 gram molecules of hydrogen, and the matter *and energy* of 1 gram molecule of oxygen, combine to the matter *and energy* of 2 gram molecules of water vapor and 293,000 joules, or units, of *free energy.*

For a hundred years the chemists thus saw only the material transformation as represented by equation (1), but overlooked and did not recognize the energy transformation coincident with the transformation of matter, though every time the experiment was made, the 293,000 J. of energy in equation (2) made themselves felt as flame, as heat and mechanical force, sometimes even explosively shattering the container in which the experiment was made. But the flame and the explosion appeared only as an incidental phenomenon without significance, as it represents and contains no part of the matter, but equation (1) gives the complete balance of matter in transformation. It was much later that the scientists realized the significance of the flame accompanying the material transformation as not a mere incidental phenomenon, but as the manifestation of the entity energy, permanent and indestructible, like matter, and the complete equation (2) appeared, giving the balance of energy as well as the balance of matter—that is, coincident with the transformation of matter is a transformation of energy, and both are indissoluble from each other, either involves the other, and both may be called different aspects of the same phenomenon.

But we have seen, when mental activity occurs in our mind, chemical and physical transformations accompany it, are coincident with it, and apparently indissoluble from it. Does there possibly exist the same relation between mental activity and the transformations of energy and matter, as we have seen to exist between the latter two? Are mental activity, energy transformation, and transformation of matter three aspects of the same biochemical phenomenon?

If for nearly a hundred years equation (1) was considered complete, until we found that one side was incomplete, and arrived at the more complete equation (2), the question may well be raised: Is equation (2) complete, dealing as it does with two entities, matter and energy, or is it not possibly still incomplete, and a third entity should appear in the equation, an entity "X," as I may call it, differing from energy and from matter, just as energy and matter differ from each other, and therefore not recognizable and measurable by the means which measure energy or matter, just as energy cannot be measured by the same means as matter?

That is, the complete equation of transformation would read:

(3) $2H_2 + O_2 = 2H_2O + 293{,}000$ J. + X, involving all three entities, matter, energy, and mind, pertaining, respectively, to the realm of chemistry, of physics, and of psychology, or possibly a broader science of which psychology is one branch.

There is no scientific evidence whatsoever of the existence of such a third entity, "X," but all our deductions have been by analogy, which proves nothing—that is, by speculation, dreaming, and unavoidably so—since in these conceptions we are close to the border line of the human mind where logical reasoning loses itself in the fog of contradiction. But at the same time there is no evidence against the conception of an entity "X"; it is not illogical, at least no more so than all such general conceptions, no more so than, for instance, that of energy or of matter. As empirical science deals with energy and matter, and entity "X" is neither, it could not be observed by any of the methods of experimental physics or chemistry.

If mind is a third entity, correlated with the entities of energy and of matter, we should expect that mental activity, or entity "X," should occur not only in the highly complex transformations of energy and of matter taking place in the brains of the highest orders of living beings, but that entity "X" should appear in all physico-chemical reactions, just as energy transformations always occur in transformations of matter, and inversely. But this seems not so, and in most of the transformations of energy and of matter entity "X" does not appear. However, we have no satisfactory means of recognizing entity "X," no methods of studying it. Therefore, it may well be that it is noticed only in those rare instances when it appears of high intensity, but in most reactions entity "X" may be so small or appear in such way as to escape observation by the means and by the methods now available. Like energy or matter, entity "X" may have many forms in which it is not recognized by us, just as for a long time the flame was not recognized as the entity energy.

To illustrate, again by analogy: In many transformations of matter, indeed, in most of the more complex ones of the organic world, the concurrent energy transformation is of such slowness and of such low intensity that it appears nonexisting, and can be discovered and measured only by the delicate experiments devised by science. Furthermore, the energy may appear in different forms. Thus the 293,000 J. of energy in equation (2) may appear as heat, or as electrical energy, or as a combination of heat, light, sound, and mechanical energy. Now assume that we could observe and notice only one of the forms of energy—for instance, only electrical energy. We should then find that in the equation (1) we only sometimes get energy—that is, electrical energy—under special peculiar conditions, but usually do not seem to get any of the entity energy, simply because we do not recognize it in the form in which it appears. Analogously, there might be a term of entity "X" in all transformations, even such simple ones as equation (3), but entity "X" may appear in a far different, simpler form. It would mean that "mind" is only one form of entity "X," perhaps the high-grade form, as it appears in highly complex reactions. In the simpler physico-chemical processes of nature, entity "X" also would appear, but in other, simpler forms. It would mean that things such as mind and intellect are not limited to the higher living beings, but characteristics akin thereto would be found grading down throughout all living and inanimate nature. This does not appear unreasonable when we consider that some characteristics of life are found throughout all nature, even in the crystal which, in its mother liquor, repairs a lesion, "heals a wound," or which, in the colloidal solution, may be "poisoned" by prussic acid.

Assume, then, that mind, intellect, personality, the ego, were forms of a third entity, an entity "X," correlated in nature with the entities energy and matter. Then, just as energy and matter continuously change their forms, so with the transformations of energy and of matter, entity "X" would continuously change, disappear in one form and reappear in another form. Entity "X" could therefore not exist permanently in one and the same form, and the permanency of the ego—that is, immortality—would still be illogical, would not exist within the realm of science, but would carry us beyond the limitations of the human mind into the unknowable. Permanency of the ego—that is, individual immortality—would require a form of entity "X," in which it is not further transformable. This would be the case if the transformations of entity "X" are not completely reversible, but tend one definite direction, from lower-grade to higher-grade forms, and the latter thus would gradually build up to increasing permanency. There is nothing unreasonable in this, but a similar condition—in the reverse direction—

exists with the transformations of energy. They also are not completely reversible, but tend in a definite direction, from higher- to lower-grade form—unavailable heat energy (the increase of entropy by the second law of thermodynamics). Thus in infinite time the universe should come to a standstill, in spite of the law of conservation of energy, by all energy becoming unavailable for further transformation—that is, becoming dead energy. If entity "X" existed, could it not also have become unavailable for further transformation by reaching its maximum high-grade form and thus become not susceptible to further change—that is, "immortal"—just as the unavailable heat of the physicist is "immortal," and not capable of further transformation? Here we are again in the fog of illogic, beyond the limitations. However, it sounds familiar to the Nirvana of the Buddhist.

Physics and chemistry obviously could not deal with entity "X," and the most delicate and sensitive physical or chemical instruments could get no indication of it, and all attempts at investigation by physical or chemical means thus must be doomed to failure. But such investigations of entity "X" belong to the realm of the science of psychology, or, rather, a broader science, of which psychology is one branch dealing with one form of entity "X," mind, just as, for instance, electro-physics is one branch of the broader science of physics, dealing with electrical energy, while physics deals with all forms of energy.

In concluding, I wish to say that nothing in the preceding speculations can possibly encourage spiritism or other pseudo-science. On the contrary, from the preceding it is obvious that the alleged manifestations of spiritism must be fake or self-deception, since they are manifestations of energy. Entity "X," if it exists, certainly is not energy, and therefore could not manifest itself as such.